GEOMETRIC ANALYSIS AROUND
SCALAR CURVATURES

LECTURE NOTES SERIES
Institute for Mathematical Sciences, National University of Singapore

Series Editors: Chitat Chong and Kwok Pui Choi
                   Institute for Mathematical Sciences
                   National University of Singapore

ISSN: 1793-0758

Published

*For the complete list of titles in this series, please go to
http://www.worldscientific.com/series/LNIMSNUS

Lecture Notes Series, Institute for Mathematical Sciences, National University of Singapore

Vol. 31

GEOMETRIC ANALYSIS AROUND SCALAR CURVATURES

Editors

Fei Han
National University of Singapore, Singapore

Xingwang Xu
National University of Singapore, Singapore

Weiping Zhang
Nankai University, China

World Scientific

NEW JERSEY · LONDON · SINGAPORE · BEIJING · SHANGHAI · HONG KONG · TAIPEI · CHENNAI · TOKYO

Published by

World Scientific Publishing Co. Pte. Ltd.

5 Toh Tuck Link, Singapore 596224

USA office: 27 Warren Street, Suite 401-402, Hackensack, NJ 07601

UK office: 57 Shelton Street, Covent Garden, London WC2H 9HE

Library of Congress Cataloging-in-Publication Data

Names: Han, Fei (Mathematician), editor. | Xu, Xingwang, 1957– editor. |
   Zhang, Weiping, 1964– editor.
Title: Geometric analysis around scalar curvatures / edited by Fei Han (NUS, Singapore),
   Xingwang Xu (NUS, Singapore), Weiping Zhang (Nankai University, China).
Other titles: Lecture notes series (National University of Singapore.
   Institute for Mathematical Sciences) ; v. 31.
Description: New Jersey : World Scientific, 2016. | Series: Lecture notes series,
   Institute for Mathematical Sciences, National University of Singapore ; volume 31 |
   Includes bibliographical references.
Identifiers: LCCN 2016003762 | ISBN 9789813100541 (hardcover : alk. paper)
Subjects: LCSH: Curvature. | Geometric analysis. | Riemannian manifolds. | Algebraic topology.
Classification: LCC QA645 .G46 2016 | DDC 516.3/62--dc23
LC record available at http://lccn.loc.gov/2016003762

British Library Cataloguing-in-Publication Data
A catalogue record for this book is available from the British Library.

CONTENTS

FOREWORD

The Institute for Mathematical Sciences (IMS) at the National University of Singapore was established on 1 July 2000. Its mission is to foster mathematical research, both fundamental and multidisciplinary, particularly research that links mathematics to other efforts of human endeavor, and to nurture the growth of mathematical talent and expertise in research scientists, as well as to serve as a platform for research interaction between scientists in Singapore and the international scientific community.

The Institute organizes thematic programs of longer duration and mathematical activities including workshops and public lectures. The program or workshop themes are selected from among areas at the forefront of current research in the mathematical sciences and their applications.

Each volume of the *IMS Lecture Notes Series* is a compendium of papers based on lectures or tutorials delivered at a program/workshop. It brings to the international research community original results or expository articles on a subject of current interest. These volumes also serve as a record of activities that took place at the IMS.

We hope that through the regular publication of these *Lecture Notes* the Institute will achieve, in part, its objective of reaching out to the community of scholars in the promotion of research in the mathematical sciences.

December 2015

Chitat Chong
Kwok Pui Choi
Series Editors

PREFACE

During November 1 to December 31, 2014, the program "Scalar Curvature in Manifold Topology and Conformal Geometry" was held at the Institute for Mathematical Sciences (IMS) in the National University of Singapore (NUS). Curvatures are central concepts in differential geometry for understanding the shape of higher dimensional curved spaces. Scalar curvature, the simplest yet fundamental, has very deep and rich connections to analysis and topology on manifolds. The purpose of the program is to bring together researchers working on relevant areas to communicate ideas, unravel the connections and stimulate possible research collaborations.

The program consisted of two international workshops and one winter school. The winter school was held during the week of 15–19 December 2014 to train graduate students and young researchers. The speakers were either top experts or outstanding young researchers in the area. This volume collects three expanded lecture notes taught in this winter school. In the following, we shall give a brief overview.

Chapter 1. *Lectures on the Fourth-Order Q Curvature Equation* by Fengbo Hang and Paul C. Yang (lecturer). In conformal geometry, a major tool is conformal covariant operators and their associated curvature invariants. The main purpose of this lecture is to introduce the fourth-order Paneitz operator and Q curvature, and their applications in understanding the conformal geometry and the topology of underlying manifold. Q-curvature is the quadratic analogue of the scalar curvature and Paneitz operator is the fourth-order type of conformal covariant operator similar to the conformal Laplace operator. These topics are recently developed and are still very active in conformal geometry.

Chapter 2. *An Introduction to the Finite and Infinite Dimensional Reduction Methods* by Manuel del Pino and Juncheng Wei (lecturer). The authors introduce the finite dimensional and infinite dimensional Liapunov–Schmidt reduction methods. These methods are illustrated by various model problems, especially in the infinite dimensional case, the Allen–Cahn equation and minimal surfaces equation. Current progress made by the authors demonstrated that these methods are very powerful in handling some nonlinear differential equations arising from geometry or physics.

Chapter 3. *Einstein Constraint Equations on Riemannian Manifolds* by Quốc Anh Ngô. In the lecture notes, Ngô first briefly introduces the Einstein equation in general relativity theory. He then discusses the initial value problem and the derivation of Einstein constraint equations. In the last part, he illustrates how to apply the interesting slice method to study the Einstein constraint equations.

We would like to thank all the participants of the program for their support and stimulating interactions during this successful two-month program.

Our program was funded by IMS. We are indebted to Professor Chi Tat Chong, Director of IMS, for his guidance throughout the program and his leadership of IMS in providing us a relaxed but inspiring environment. Thanks also go to the IMS staff, especially to Emily Chan Ee Cheng and Eileen Tan whom, without their wonderful administrative support, the workshops would not have been as successful.

November 2015   Fei Han
                                National University of Singapore
                                                    Singapore

                                                 Xingwang Xu
                                National University of Singapore
                                                    Singapore

                                               Weiping Zhang
                                      Nankai University, China

                                              Volume Editors

LECTURES ON THE FOURTH-ORDER Q CURVATURE EQUATION

Fengbo Hang

Courant Institute
New York University
251 Mercer Street
New York, NY 10012 USA
fengbo@cims.nyu.edu

Paul C. Yang

Department of Mathematics
Princeton University
Fine Hall, Washington Road
Princeton, NJ 08544 USA
yang@math.princeton.edu

We discuss some open problems and recent progress related to the fourth-order Paneitz operator and Q curvature in dimensions other than 4.

Contents

1. Introduction

In conformal geometry, a major tool is a family of conformal covariant operators and their associated curvature invariants. In dimension $n > 2$, the conformal Laplacian operator

$$L = -\frac{4(n-1)}{n-2}\Delta + R \tag{1.1}$$

enjoys the following covariance property,

$$L_{\rho^{\frac{4}{n-2}}g}\varphi = \rho^{-\frac{n+2}{n-2}}L_g(\rho\varphi) \tag{1.2}$$

for any smooth positive function ρ (see [31]). Here R denotes the scalar curvature. The associated transformation law of scalar curvature follows,

$$R_{\rho^{\frac{4}{n-2}}g} = L_{\rho^{\frac{4}{n-2}}g}1 = \rho^{-\frac{n+2}{n-2}}L_g\rho. \tag{1.3}$$

A fundamental result is the solution of the Yamabe problem [3, 40, 42, 47], which is related to the sharp constant of the associated Sobolev inequality. Since then, there is a large literature on the analysis and geometry of this equation. In order to gain additional information on the Ricci tensor, the fourth-order Q curvature equation comes into play.

Let (M, g) be a smooth Riemannian manifold with dimension $n \geq 3$, the Q curvature is given by ([4, 36])

$$Q = -\frac{1}{2(n-1)}\Delta R - \frac{2}{(n-2)^2}|Rc|^2 + \frac{n^3 - 4n^2 + 16n - 16}{8(n-1)^2(n-2)^2}R^2 \tag{1.4}$$

$$= -\Delta J - 2|A|^2 + \frac{n}{2}J^2.$$

Here Rc is the Ricci tensor and

$$J = \frac{R}{2(n-1)}, \quad A = \frac{1}{n-2}(Rc - Jg). \tag{1.5}$$

The Paneitz operator is defined as

$$P\varphi \tag{1.6}$$

$$= \Delta^2\varphi + \frac{4}{n-2}div(Rc(\nabla\varphi, e_i)e_i) - \frac{n^2 - 4n + 8}{2(n-1)(n-2)}div(R\nabla\varphi) + \frac{n-4}{2}Q\varphi$$

$$= \Delta^2\varphi + div(4A(\nabla\varphi, e_i)e_i - (n-2)J\nabla\varphi) + \frac{n-4}{2}Q\varphi.$$

Here $e_1, \ldots, e_n$ is a local orthonormal frame with respect to g. Note that the use of J and A (Schouten tensor) simplifies the formulas of Q curvature and Paneitz operator.

In dimension $n \neq 4$, the operator satisfies

$$P_{\rho^{\frac{4}{n-4}}g}\varphi = \rho^{-\frac{n+4}{n-4}}P_g(\rho\varphi) \tag{1.7}$$

for any positive smooth function ρ. This is similar to (1.2). As a consequence we have

$$Q_{\rho^{\frac{4}{n-4}}g} = \frac{2}{n-4}P_{\rho^{\frac{4}{n-4}}g}1 = \frac{2}{n-4}\rho^{-\frac{n+4}{n-4}}P_g\rho. \tag{1.8}$$

In dimension 4, the Paneitz operator satisfies

$$P_{e^{2w}g}\varphi = e^{-4w}P_g\varphi \tag{1.9}$$

and the Q curvature transforms as

$$Q_{e^{2w}g} = e^{-4w}\left(P_g w + Q_g\right). \tag{1.10}$$

This should be compared to the conformal invariance of $-\Delta$ on surface and the transformation law of Gaussian curvature under a conformal change of metric.

The main theme of research is to find out the role of Paneitz operator and Q curvature in understanding the geometry of a conformal class and the topology of underlying manifold. For example, we would like to know how the spectral property of Paneitz operator affects the topology. Below we will start with dimension 4, when the Q curvature equation and its applications is relatively well understood. Then we will discuss recent progress in dimension $n \geq 5$ about the Green's function of Paneitz operator and the solution to finding constant Q curvature in a fixed conformal class. At last we will turn to the dimension 3, where the Q curvature equation is particularly intriguing and of very different nature from the scalar curvature equation. Open problems will be pointed out along the way.

2. Dimension 4

A basic fact that makes the Q curvature interesting is its appearance in the Chern-Gauss-Bonnet formula. For a closed 4-manifold (M, g) we have

$$\int_M Q d\mu + \frac{1}{4}\int_M |W|^2 d\mu = 8\pi^2 \chi(M). \tag{2.1}$$

Here W is the Weyl tensor. It follows from the pointwise conformal invariance of $|W|^2 d\mu$ and (2.1) that the Q curvature integral is a global conformal invariant which we denote by κ_g i.e.

$$\kappa_g = \int_M Q_g d\mu_g \tag{2.2}$$

and $\kappa_{\tilde{g}} = \kappa_g$ for any $\tilde{g} \in [g]$, the conformal class of g. A basic result about this invariant is the following sharp upper bound:

Theorem 2.1: [16] *Let (M, g) be a smooth compact four manifold. If $L_g > 0$, then $\kappa_g \leq 16\pi^2$ with equality holds if and only if (M, g) is conformal diffeomorphic to the standard four sphere.*

Theorem 2.1 follows from an identity found in [25]. The identity will have a crucial counterpart in other dimensions.

Theorem 2.2: [25] *Let (M, g) be a 4-dimensional smooth compact Riemannian manifold with $L_g > 0$. For $p \in M$, let $G_{L,p}$ be the Green's function for L_g with pole at p, then we have $\left| Rc_{G_{L,p}^2 g} \right|_g^2$ is bounded and*

$$P\left(\log G_{L,p}\right) = 16\pi^2 \delta_p - \frac{1}{2} \left| Rc_{G_{L,p}^2 g} \right|_g^2 - Q \tag{2.3}$$

in distribution sense.

Choosing 1 as test function in (2.3) we see

$$\int_M Q d\mu = 16\pi^2 - \frac{1}{2} \int_M \left| Rc_{G_{L,p}^2 g} \right|_g^2 d\mu \le 16\pi^2.$$

If equality holds, then $Rc_{G_{L,p}^2 g} = 0$ and by the relative volume comparison theorem we conclude (M, g) must be conformal equivalent to the standard S^4 (see Section 5 of [25]).

To study the Q curvature equation, it is important that the Paneitz operator be nonnegative with only constant functions in its kernel. A quite general condition ensuring such kind of positivity is given by

Theorem 2.3: [16] *Let (M, g) be a smooth compact 4-dimensional Riemannian manifold with $L_g > 0$ and $\kappa_g \ge 0$, then the Paneitz operator $P \ge 0$ and the kernel of P consists of constant functions.*

As an application of Theorems 2.1 and 2.3, we have a general existence result for a conformal metric of constant Q curvature. This is analogous to the existence of constant Gauss curvature metrics in dimension two. Let us consider the following functionals

$$I(w) = \int_M |W|^2 \, w d\mu - \frac{1}{4} \left(\int_M |W|^2 \, d\mu \right) \log \left(\frac{1}{\mu(M)} \int_M e^{4w} d\mu \right), \tag{2.4}$$

$$II(w) \tag{2.5}$$
$$= \int_M Pw \cdot w d\mu + 2 \int_M Q w d\mu - \frac{1}{2} \left(\int_M Q d\mu \right) \log \left(\frac{1}{\mu(M)} \int_M e^{4w} d\mu \right),$$

and

$$III(w) = \int_M J_{e^{2w} g}^2 d\mu_{e^{2w} g} - \int_M J^2 d\mu. \tag{2.6}$$

The Euler-Lagrange equation of functional II is given by

$$Pw + Q - \frac{\int_M Q d\mu}{\int_M e^{4w} d\mu} e^{4w} = 0. \tag{2.7}$$

Or in other words,

$$Q_{e^{2w}g} = const. \tag{2.8}$$

On the other hand, the Euler-Lagrange equation for functional III is

$$-\Delta_{e^{2w}g} J_{e^{2w}g} = 0. \tag{2.9}$$

In [10] the general functional $F = \gamma_1 I + \gamma_2 II + \gamma_3 III$ was studied.

Theorem 2.4: [10] *If the functional F satisfies*

$$\gamma_2 > 0, \quad \gamma_3 > 0 \tag{2.10}$$

and

$$\kappa = \frac{\gamma_1}{2} \int_M |W|^2 d\mu + \gamma_2 \int_M Q d\mu < 16\pi^2 \gamma_2, \tag{2.11}$$

then there exists a minimizer for

$$\inf_{w \in H^2(M)} F(w). \tag{2.12}$$

Any minimizer must be smooth. If w is a minimizer and we write $\tilde{g} = e^{2w}g$, then

$$\frac{\gamma_1}{2} \left|\widetilde{W}\right|^2_{\tilde{g}} + \gamma_2 \tilde{Q} - \gamma_3 \tilde{\Delta}\tilde{J} = \frac{\kappa}{\tilde{\mu}(M)}. \tag{2.13}$$

Moreover for any $\varphi \in H^2(M)$ with

$$\int_M \varphi d\tilde{\mu} = 0, \tag{2.14}$$

we have

$$\gamma_2 \int_M \widetilde{P}\varphi \cdot \varphi d\tilde{\mu} + \gamma_3 \int_M \left[\left(\tilde{\Delta}\varphi + \left|\tilde{\nabla}\varphi\right|^2_{\tilde{g}}\right)^2 - 2\tilde{J}\left|\tilde{\nabla}\varphi\right|^2_{\tilde{g}}\right] d\tilde{\mu} \tag{2.15}$$
$$\geq \frac{\kappa}{2} \log\left(\frac{1}{\tilde{\mu}(M)} \int_M e^{4\varphi} d\tilde{\mu}\right).$$

Here $\int_M \widetilde{P}\varphi \cdot \varphi d\tilde{\mu}$ is understood in distribution sense.

For the functional II, we have a similar existence result.

Theorem 2.5: [10] *If*

$$\kappa_g = \int_M Q d\mu < 16\pi^2, \tag{2.16}$$

$P \geq 0$ *and the kernel of P consists only of constant functions, then*

$$\inf_{w \in H^2(M)} II(w) \tag{2.17}$$

is achieved. Any minimizer must be smooth. If w is a minimizer and we write $\tilde{g} = e^{2w}g$, then

$$\tilde{Q} = \frac{\kappa_g}{\tilde{\mu}(M)}. \tag{2.18}$$

Moreover for any $\varphi \in H^2(M)$ with

$$\int_M \varphi d\tilde{\mu} = 0, \tag{2.19}$$

we have

$$\int_M \tilde{P}\varphi \cdot \varphi d\tilde{\mu} \geq \frac{\kappa_g}{2} \log\left(\frac{1}{\tilde{\mu}(M)} \int_M e^{4\varphi} d\tilde{\mu}\right). \tag{2.20}$$

Here $\int_M \tilde{P}\varphi \cdot \varphi d\tilde{\mu}$ is understood in distribution sense.

More results on the existence of conformal metrics with constant Q curvature can be found in [12]. The main ingredient for Theorems 2.4 and 2.5 is the following version of Adams inequality ([1]):

Theorem 2.6: [5, 13] *Let (M, g) be a smooth compact 4-dimensional Riemannian manifold with $P \geq 0$ and kernel of P consists only of constant functions, then for any $w \in H^2(M)$ with*

$$\int_M w d\mu = 0, \tag{2.21}$$

we have

$$\int_M \exp\left(32\pi^2 \frac{w^2}{\int_M Pw \cdot w d\mu}\right) d\mu \leq c(M, g) < \infty. \tag{2.22}$$

In particular

$$\log\left(\frac{1}{\mu(M)} \int_M e^{4w} d\mu\right) \leq \frac{1}{8\pi^2} \int_M Pw \cdot w d\mu + c(M, g). \tag{2.23}$$

Here $\int_M Pw \cdot w d\mu$ is understood in distribution sense.

Adams inequality was discovered in [1] with the motivation of simplifying the original proof in [34]. In particular, a higher order sharp inequality was derived through the O'Neil inequality for convolution operator (see [35]) and a 1-dimensional calculus lemma due to Adams-Garsia. Theorem 2.6 can be proven by modifying O'Neil inequality and the calculus lemma.

For some geometrical and topological applications of these related equations, we refer the readers to [7, 8, 15].

3. Dimension of at Least 5

The analysis of Q curvature and Paneitz operator in dimension greater than 4 has some similarity to the analysis of scalar curvature and conformal Laplacian operator in dimension greater than 2. The research related to Yamabe problem serves as a nice model for asking interesting questions in the study of Paneitz operator. However, due to the fact second-order differential equations are much better understood than higher order differential equations, sometimes the analogous problem for Q curvature can be more challenging.

Based on the fact the first eigenfunction of conformal Laplacian operator can always be chosen as positive everywhere, it was observed in [30] that in a fixed conformal class, we can always find a metric whose scalar curvature is only of one sign i.e. the scalar curvature is either strictly positive, or identically zero, or strictly negative.

Problem 3.1: *Let (M, g) be a smooth compact Riemannian manifold with dimension $n \geq 5$, can we always find a conformal metric $\widetilde{g}$ such that $\widetilde{Q}$ is either strictly positive, or identically zero, or strictly negative?*

This seems to be a difficult question. One of the obstacle is fourth-order symmetric elliptic operators can have no positive first eigenfunction at all. Indeed let M be any smooth compact Riemannian manifold, λ be the smallest positive eigenvalue of $-\Delta$, then the first eigenfunction of $(-\Delta)^2 + 2\lambda\Delta$ must change sign. Though the answer to Problem 3.1 remains mysterious, partial solution to a related problem was found recently in [25]. Recall on a smooth compact Riemannian manifold (M, g) with dimension greater than 2, we have

$$\exists \widetilde{g} \in [g] \text{ with } \widetilde{R} > 0 \Longleftrightarrow \lambda_1 (L_g) > 0.$$

Here $[g]$ denotes the conformal class of metrics associated with g. The same statement remains true if we replace ">" by "<" or "=" (see [31]). It is worth pointing out the sign of $\lambda_1 (L_g)$ is a conformal invariant. In particular, the above statement gives a conformal invariant condition which is equivalent to the existence of a conformal metric with positive scalar curvature.

Problem 3.2: *Let (M, g) be a smooth compact Riemannian manifold with dimension $n \geq 5$, can we find a conformal invariant condition which is equivalent to the existence of a conformal metric with positive Q curvature? Same questions can be asked when "positive" is replaced by "negative" or "zero".*

[25] gives a partial answer to this problem under the assumption the Yamabe invariant $Y(g) > 0$.

Theorem 3.3: [25] *Let $n \geq 5$ and (M^n, g) be a smooth compact Riemannian manifold with Yamabe invariant $Y(g) > 0$, then the following statements are equivalent:*

(1) $\exists \tilde{g} \in [g]$ with $\tilde{Q} > 0$.
(2) $\ker P_g = 0$ and the Green's function of Paneitz operator $G_P(p, q) > 0$ for any $p, q \in M, p \neq q$.
(3) $\ker P_g = 0$ and there exists a $p \in M$ such that $G_P(p, q) > 0$ for $q \in M \setminus \{p\}$.

By transformation law (1.7), we know $\ker P_g = 0$ is a conformal invariant condition. Moreover under this assumption, the Green's functions of Paneitz operator G_P satisfy

$$G_{P, \rho^{\frac{4}{n-4}} g}(p, q) = \rho(p)^{-1} \rho(q)^{-1} G_{P, g}(p, q). \tag{3.1}$$

In particular, the fact $G_P > 0$ is also a conformal invariant condition. Of course this condition is clearly more complicated than the one given for the scalar curvature case, however the main strength of Theorem 3.3 lies in that it gives an easy to check necessary and sufficient condition for the positivity of the Green's function of Paneitz operator for metrics of positive Yamabe class. As we will see shortly, the positivity of Green's function is crucial in the study of Q curvature equation.

The main ingredients in proof of Theorem 3.3 is an identity similar to (2.3) in higher dimension.

Theorem 3.4: [25] *Assume $n \geq 5$, (M^n, g) is a smooth compact Riemannian manifold with $Y(g) > 0$, $p \in M$, then we have $G_{L,p}^{\frac{n-4}{n-2}} \left| Rc_{G_{L,p}^{\frac{4}{n-2}} g} \right|^2_g \in$*

$L^1(M)$ *and*

$$P\left(G_{L,p}^{\frac{n-4}{n-2}}\right) = c_n \delta_p - \frac{n-4}{(n-2)^2} G_{L,p}^{\frac{n-4}{n-2}} \left| Rc_{G_{L,p}^{\frac{4}{n-2}} g} \right|_g^2 \tag{3.2}$$

in distribution sense. Here

$$c_n = 2^{-\frac{n-6}{n-2}} n^{\frac{2}{n-2}} (n-1)^{-\frac{n-4}{n-2}} (n-2)(n-4) \omega_n^{\frac{2}{n-2}}, \tag{3.3}$$

ω_n *is the volume of unit ball in* $\mathbb{R}^n$, $G_{L,p}$ *is the Green's function of conformal Laplacian operator with pole at* p.

Here we will give another conformal invariant condition for the existence of conformal metric with positive Q curvature. To achieve this, we first introduce some notations.

Let (M, g) be a smooth compact Riemannian manifold. If $K = K(p, q)$ is a suitable function on $M \times M$, we define an operator T_K as

$$T_K(\varphi)(p) = \int_M K(p, q) \varphi(q) d\mu(q) \tag{3.4}$$

for any nice function φ on M. If $K' = K'(p, q)$ is another function on $M \times M$, then we write

$$(K * K')(p, q) = \int_M K(p, s) K'(s, q) d\mu(s). \tag{3.5}$$

If $n \geq 5$ and $Y(g) > 0$, we write

$$H(p, q) \tag{3.6}$$
$$= 2^{\frac{n-6}{n-2}} n^{-\frac{2}{n-2}} (n-1)^{\frac{n-4}{n-2}} (n-2)^{-1} (n-4)^{-1} \omega_n^{-\frac{2}{n-2}} G_L(p, q)^{\frac{n-4}{n-2}},$$

and

$$\Gamma_1(p, q) \tag{3.7}$$
$$= 2^{\frac{n-6}{n-2}} n^{-\frac{2}{n-2}} (n-1)^{\frac{n-4}{n-2}} (n-2)^{-3} \omega_n^{-\frac{2}{n-2}} G_L(p, q)^{\frac{n-4}{n-2}} \left| Rc_{G_{L,p}^{\frac{4}{n-2}} g} \right|_g^2 (q).$$

Then (3.2) becomes

$$P_q H(p, q) = \delta_p(q) - \Gamma_1(p, q). \tag{3.8}$$

Note that by the calculation in Section 2 of [25],

$$\Gamma_1(p, q) = O\left(\overline{pq}^{4-n}\right), \tag{3.9}$$

here $\overline{pq}$ denotes the distance between p and q. Assume for all $p \in M$,

$$0 \le \int_M \Gamma_1 (p,q) \, d\mu (q) \le \alpha < \infty, \tag{3.10}$$

then

$$\|T_{\Gamma_1}\varphi\|_{L^\infty(M)} \le \alpha \|\varphi\|_{L^\infty(M)}. \tag{3.11}$$

Moreover, if we let $\widetilde{g} = \rho^{\frac{4}{n-4}} g$, here ρ is a positive smooth function, then for any smooth function φ on M,

$$T_{\widetilde{\Gamma}_1} (\varphi) = \rho^{-1} T_{\Gamma_1} (\rho\varphi). \tag{3.12}$$

In other words, $T_{\widetilde{\Gamma}_1}$ is similar to T_{Γ_1}. Hence they have the same spectrum and spectral radius i.e. $\sigma \left(T_{\widetilde{\Gamma}_1} \right) = \sigma \left(T_{\Gamma_1} \right)$ and $r_\sigma \left(T_{\widetilde{\Gamma}_1} \right) = r_\sigma \left(T_{\Gamma_1} \right)$ (the spectral radius).

Theorem 3.5: *Assume $n \ge 5$, (M^n, g) is a smooth compact Riemannian manifold with $Y(g) > 0$, then*

$$\exists \widetilde{g} \in [g] \text{ with } \widetilde{Q} > 0. \iff \text{ the spectral radius } r_\sigma (T_{\Gamma_1}) < 1.$$

Moreover if $r_\sigma (T_{\Gamma_1}) < 1$, then $\ker P = 0$ and

$$G_P = H + \sum_{k=1}^{\infty} \Gamma_k * H, \tag{3.13}$$

here

$$\Gamma_k = \Gamma_1 * \cdots * \Gamma_1 \text{ (k times)}, \tag{3.14}$$

H and Γ_1 are given in (3.6) and (3.7). The convergence in (3.13) is uniform in the sense that

$$G_P - H - \sum_{k=1}^{l} \Gamma_k * H \longrightarrow 0$$

uniformly on $M \times M$ as $l \to \infty$. In particular, $G_P \ge H$, moreover if $G_P (p,q) = H (p,q)$ for some $p \ne q$, then (M, g) is conformal equivalent to the standard S^n.

Proof: Assume there exists a $\widetilde{g} \in [g]$ with $\widetilde{Q} > 0$, then we hope to show $r_\sigma (T_{\Gamma_1}) < 1$. Because $r_\sigma (T_{\Gamma_1}) = r_\sigma \left(T_{\widetilde{\Gamma}_1} \right)$, replacing g with $\widetilde{g}$ we can assume the background metric satisfies $Q > 0$. By (3.8) we know for any smooth function φ,

$$\varphi = T_H (P\varphi) + T_{\Gamma_1} (\varphi). \tag{3.15}$$

Taking $\varphi = 1$ in (3.15) we get

$$\int_M \Gamma_1(p, q)\, d\mu(q) = 1 - \frac{n-4}{2} \cdot \int_M H(p, q)\, Q(q)\, d\mu(q). \qquad (3.16)$$

Using the fact $Q > 0$ we know there exists a constant α such that

$$\int_M \Gamma_1(p, q)\, d\mu(q) \le \alpha < 1$$

for all $p \in M$. It follows that

$$\|T_{\Gamma_1}\|_{\mathcal{L}(L^\infty, L^\infty)} \le \alpha$$

and hence

$$r_\sigma(T_{\Gamma_1}) \le \alpha < 1.$$

On the other hand, assume $r_\sigma(T_{\Gamma_1}) < \alpha < 1$, then we can find a constant k_0 such that for $k \ge k_0$,

$$\|T_{\Gamma_k}\|_{\mathcal{L}(L^\infty, L^\infty)} < \alpha^k.$$

It follows that

$$\int_M \Gamma_k(p, q)\, d\mu(q) < \alpha^k.$$

Fix $m > \frac{n}{4}$, using estimate (3.9) we see for all $k \ge k_0 + m$,

$$\|\Gamma_k\|_{L^\infty} \le \alpha^{k-m} \|\Gamma_m\|_{L^\infty} \le c\alpha^k.$$

In particular, $\|\Gamma_k\|_{L^\infty} \to 0$ and

$$\|\Gamma_k * H\|_{L^\infty} \le c\|\Gamma_k\|_{L^\infty} \le c\alpha^k.$$

Iterating (3.15), we see

$$\varphi = T_{H + \Gamma_1 * H + \cdots + \Gamma_{k-1} * H}(P\varphi) + T_{\Gamma_k}(\varphi).$$

Let $k \to \infty$, we see

$$\varphi = T_{H + \sum_{k=1}^\infty \Gamma_k * H}(P\varphi).$$

In particular, $P\varphi = 0$ implies $\varphi = 0$ i.e. $\ker P = 0$. Moreover,

$$G_P = H + \sum_{k=1}^\infty \Gamma_k * H.$$

In particular, $G_P \ge H > 0$. If $G_P(p, q) = H(p, q)$ for some $p \ne q$, then $\Gamma_1(p, \cdot) = 0$, in other words

$$Rc_{G_{L,p}^{\frac{4}{n-2}} g} = 0.$$

Since $\left(M \backslash \{p\}, G_{L,p}^{\frac{4}{n-2}} g \right)$ is asymptotically flat, it follows from relative vol-

ume comparison theorem that $\left(M \backslash \{p\}, G_{L,p}^{\frac{4}{n-2}} g \right)$ is isometric to $\mathbb{R}^n$, hence (M,g) is conformal equivalent to standard S^n.

Since $G_P > 0$, it follows from Theorem 3.3 that there exists $\tilde{g} \in [g]$ with $\tilde{Q} > 0$.   □

We remark that the infinite series expansion of G_P in (3.13) is similar to those for Green's function of Laplacian in [2].

Remark 3.6: Indeed it follows from (3.16) that as long as $Y(g) > 0$ and

$$\int_M H(p,q) Q(q) \, d\mu(q) > 0$$

for all $p \in M$, then $r_\sigma(T_{\Gamma_1}) < 1$. In particular, this is the case when $Q \geq 0$ and not identically zero.

Problem 3.7: *Let (M,g) be a smooth compact Riemannian manifold with dimension $n \geq 5$, can we find a metric $\tilde{g} \in [g]$ such that $\tilde{Q} = const$?*

This turns out to be a difficult problem with only partial solutions available. If we write the unknown metric $\tilde{g} = \rho^{\frac{4}{n-4}} g$, then we need to solve

$$P\rho = const \cdot \rho^{\frac{n+4}{n-4}}, \quad \rho \in C^\infty(M), \rho > 0. \tag{3.17}$$

As in the case of Yamabe problem, (3.17) has a variational structure. Indeed, for $u \in C^\infty(M)$, let

$$E(u) = \int_M Pu \cdot u \, d\mu \tag{3.18}$$

$$= \int_M \left[(\Delta u)^2 - 4A(\nabla u, \nabla u) + (n-2)J|\nabla u|^2 + \frac{n-4}{2} Q u^2 \right] d\mu.$$

Clearly, we can extend $E(u)$ continuously to $u \in H^2(M)$. Let

$$Y_4(g) = \inf_{u \in H^2(M) \backslash \{0\}} \frac{E(u)}{\|u\|_{L^{\frac{2n}{n-4}}}^2}, \tag{3.19}$$

then $Y_4(g)$ is a conformal invariant in the same spirit as $Y(g)$. If $Y_4(g)$ is achieved at a smooth positive function ρ, then it satisfies (3.17). On the other hand, even if $Y_4(g)$ is achieved at a function $u \in H^2(M)$, we cannot conclude whether u changes sign or not. An observation made in [39] says that if $P > 0$ and $G_P > 0$, then the minimizer must be smooth and either strictly positive or strictly negative. We remark that it had been observed in [27, 28, 29] that the positivity of Green's function of Paneitz operator plays crucial roles in various issues related to Q curvature. Without the classical maximum principle, it is hard to know the sign of Green's function of the fourth-order operator. A breakthrough was made in [18], which provides an easy to check sufficient condition for the positivity of Green's function.

Theorem 3.8: [18] *Assume $n \geq 5$, (M^n, g) is a smooth compact Riemannian manifold with $R > 0$, $Q \geq 0$ and not identically zero, then $P > 0$. Moreover if u is a nonzero smooth function with $Pu \geq 0$, then $u > 0$ and $R_{u^{\frac{4}{n-4}}g} > 0$. In particular, $G_P > 0$.*

Note that the necessary and sufficient condition in Theorem 3.3 is motivated by [18, 29]. The final solution of Yamabe problem uses the positive mass theorem (see [31, 40]). The corresponding statement for the Paneitz operator is established in [18, 29]. Indeed an elementary but ingenious calculation in [29] justifies the positivity of mass under the assumption of positivity of Green's function of Paneitz operator for locally conformally flat manifolds. As pointed out in [18], the same calculation carries through to nonlocally conformally flat manifolds in dimensions 5, 6 and 7 as well. A close connection between the positive mass result and formula (3.2) is found in Section 6 of [25]. Combine these with Theorems 3.3 and 3.5, we have

Theorem 3.9: [18, 25, 26, 29] *Assume $n \geq 5$, (M^n, g) is a smooth compact Riemannian manifold with $Y(g) > 0$ and the spectral radius $r_\sigma(T_{\Gamma_1}) < 1$ (Γ_1 is given by (3.7)). If $n = 5, 6, 7$ or (M, g) is locally conformally flat near $p \in M$, then $\ker P = 0$ and under conformal normal coordinate at p, $x_1, \ldots, x_n$,*

$$G_{P,p} = \frac{1}{2n(n-2)(n-4)\omega_n} \left(r^{4-n} + A + O(r) \right),$$

with the constant $A \geq 0$, here $r = |x|$, ω_n is the volume of the unit ball in $\mathbb{R}^n$. Moreover $A = 0$ if and only if (M, g) is conformal equivalent to S^n.

Indeed following Section 6 of [25] we note that under the assumption of Theorem 3.9 (see [31])

$$G_{L,p} = \frac{1}{4n(n-1)\omega_n} \left(r^{2-n} + O(r^{-1}) \right).$$

Let $H_p(q) = H(p, q)$, then

$$G_{P,p} - H_p = \frac{A}{2n(n-2)(n-4)\omega_n} + O(r).$$

It follows from (3.8) that

$$P\left(G_{P,p} - H_p\right)(q) = \Gamma_1\left(p, q\right).$$

Hence

$$
\begin{aligned}
&A \\
&= 2n(n-2)(n-4)\omega_n \int_M G_P(p,q)\Gamma_1(p,q)d\mu(q) \\
&= 2^{\frac{2(n-4)}{n-2}} n^{\frac{n-4}{n-2}}(n-1)^{\frac{n-4}{n-2}}(n-2)^{-2}(n-4)\,\omega_n^{\frac{n-4}{n-2}} \int_M G_{P,p} G_{L,p}^{\frac{n-4}{n-2}} \left| Rc_{G_{L,p}^{\frac{4}{n-2}}g} \right|_g^2 d\mu.
\end{aligned}
$$

This is exactly the formula proven in [29]. Theorem 3.9 follows from this calculation. With Theorems 3.3, 3.5 and 3.9 at hand, we are able to give the first partial solution to Problem 3.7.

Theorem 3.10: [18, 26] *Let (M, g) be a smooth compact n-dimensional Riemannian manifold with $n \geq 5$, $Y(g) > 0$, $Y_4(g) > 0$, $r_\sigma(T_{\Gamma_1}) < 1$, then*

(1) $Y_4(g) \leq Y_4(S^n)$, and equality holds if and only if (M, g) is conformally diffeomorphic to the standard sphere.

(2) $Y_4(g)$ is always achieved. Any minimizer must be smooth and cannot change sign. In particular, we can find a constant Q curvature metric in the conformal class.

(3) If (M, g) is not conformally diffeomorphic to the standard sphere, then the set of all minimizers u for $Y_4(g)$, after normalizing with $\|u\|_{L^{\frac{2n}{n-4}}} = 1$, is compact in C^∞ topology.

It is worth pointing out that for a locally conformally flat manifold with positive Yamabe invariant and Poincare exponent less than $\frac{n-4}{2}$ (see [41]), Theorem 3.10 was proved in [38] by *a priori* estimates (using method of moving planes for integral equations developed in [11]) and connecting the equation to Yamabe equation through a path of integral equations.

Note that $Y_4(g) > 0$ is the same thing as $P > 0$. Either one of the following conditions guarantee the positivity of Paneitz operator

- [18, 45]: $n \geq 5$, $R > 0$, $Q \geq 0$ and not identically zero;
- Theorem 1.6 in [9]: $n \geq 5$, $J \geq 0$, $\sigma_2(A) \geq 0$ and (M, g) is not Ricci flat.

In applications we are usually interested in metrics not just with $Q > 0$, but with both $R > 0$ and $Q > 0$. This leads us to a question similar to Problem 3.2.

Problem 3.11: *Problem 1.1 in [17]. For a smooth compact Riemannian manifold with dimension of at least 5, can we find a conformal invariant condition which is equivalent to the existence of a conformal metric with positive scalar and Q curvature?*

Theorem 3.12: *[17] Let (M, g) be a smooth compact Riemannian manifold with dimension $n \geq 6$. Denote*

$$Y_4^+(g) = \frac{n-4}{2} \inf_{\widetilde{g} \in [g]} \frac{\int_M \widetilde{Q} d\widetilde{\mu}}{(\widetilde{\mu}(M))^{\frac{n-4}{n}}} = \inf_{\substack{u \in C^\infty(M) \\ u > 0}} \frac{\int_M Pu \cdot u d\mu}{\|u\|_{L^{\frac{2n}{n-4}}}^2}$$

and

$$Y_4^*(g) = \frac{n-4}{2} \inf_{\substack{\widetilde{g} \in [g] \\ \widetilde{R} > 0}} \frac{\int_M \widetilde{Q} d\widetilde{\mu}}{(\widetilde{\mu}(M))^{\frac{n-4}{n}}}.$$

If $Y(g) > 0$ and $Y_4^*(g) > 0$, then there exists a metric $\widetilde{g} \in [g]$ satisfying $\widetilde{R} > 0$ and $\widetilde{Q} > 0$. In particular, $P > 0$, the Green's function $G_P > 0$, and $Y_4(g)$ is achieved at a positive smooth function u with $R_{u^{\frac{4}{n-4}}g} > 0$ and $Q_{u^{\frac{4}{n-4}}g} = const$. Moreover,

$$Y_4(g) = Y_4^+(g) = Y_4^*(g).$$

Corollary 3.13: *[17] Let (M, g) be a smooth compact Riemannian manifold with dimension $n \geq 6$. Then the following statements are equivalent:*

(1) $Y(g) > 0, P > 0$.
(2) $Y(g) > 0, Y_4^*(g) > 0$.
(3) there exists a metric $\widetilde{g} \in [g]$ satisfying $\widetilde{R} > 0$ and $\widetilde{Q} > 0$.

Corollary 3.13 answers Problem 3.11 for dimension of at least 6. It also tells us in Theorem 3.10, condition $r_\sigma(T_{\Gamma_1}) < 1$ is implied by the positivity of $Y(g)$ and $Y_4(g)$ when $n \geq 6$. The case $n = 5$ still remains open for Problem 3.11.

Problem 3.14: *Let (M, g) be a smooth compact Riemannian manifold with dimension $n \geq 5$, do we have*

$$Y(g) > 0, Q > 0 \Longrightarrow P > 0?$$

The answer is probably negative.

This seems to be a subtle question. Indeed from [18, 45], we know when both R and Q are positive, then P is positive definite. If we have $Y(g) > 0$ and $Q > 0$ instead, then some conformal metrics have positive scalar curvature. However, the set of metrics with positive scalar curvature may be disjoint with those with positive Q curvature. Nevertheless, Theorem 3.3 tells us $\ker P = 0$ and $G_P > 0$. In [26], it is shown this is enough to find a constant Q curvature in the conformal class. Together with Theorem 3.5, we have another partial answer to Problem 3.7.

Theorem 3.15: [26] *Let (M, g) be a smooth compact n dimensional Riemannian manifold with $n \geq 5$, $Y(g) > 0$, $r_\sigma(T_{\Gamma_1}) < 1$, then $\ker P = 0$, the Green's function of P is positive and there exists a conformal metric $\tilde{g}$ with $\tilde{Q} = 1$.*

Note that if the answer to Problem 3.14 is positive, then Theorem 3.15 would follow from Theorem 3.10. Without knowing the positivity of Paneitz operator, we cannot use the minimization problem (3.19) to find the constant Q curvature metrics. A different approach was developed in [26]. Under the assumption of Theorem 3.15, it follows from Theorem 3.5 that $\ker P = 0$ and $G_P > 0$. If we denote $f = \rho^{\frac{n+4}{n-4}}$, then equation (3.17) becomes

$$T_{G_P} f = \frac{2}{n-4} f^{\frac{n-4}{n+4}}, \quad f \in C^\infty(M), f > 0. \tag{3.20}$$

Let

$$\Theta_4(g) = \sup_{f \in L^{\frac{2n}{n+4}}(M) \setminus \{0\}} \frac{\int_M T_{G_P} f \cdot f d\mu}{\|f\|^2_{L^{\frac{2n}{n+4}}}}. \tag{3.21}$$

By (3.1), we know $\Theta_4(g)$ is a conformal invariant, moreover it has a nice geometrical description, which is local, (see Section 2.1 of [26])

$$\Theta_4(g) = \frac{2}{n-4} \sup \left\{ \frac{\int_M \tilde{Q} d\tilde{\mu}}{\|\tilde{Q}\|^2_{L^{\frac{2n}{n+4}}(M, d\tilde{\mu})}} : \tilde{g} \in [g] \right\} \tag{3.22}$$

$$= \sup_{u \in W^{4, \frac{2n}{n+4}}(M) \setminus \{0\}} \frac{\int_M Pu \cdot u d\mu}{\|Pu\|^2_{L^{\frac{2n}{n+4}}}}.$$

It follows from the classical Hardy-Littlewood-Sobolev inequality $\Theta_4(g)$ is always finite. The benefit of this formulation is if $\Theta_4(g)$ is achieved by a maximizer f, we deduce easily from the positivity of G_P that f cannot change sign. With Theorems 3.3, 3.5 and 3.9 at hand, we have the following statement about extremal problem for $\Theta_4(g)$:

Theorem 3.16: [26] *Assume (M,g) is a smooth compact n-dimensional Riemannian manifold with $n \geq 5$, $Y(g) > 0$, $r_\sigma(T_{\Gamma_1}) < 1$, then*

(1) $\Theta_4(g) \geq \Theta_4(S^n)$, *here S^n has the standard metric. $\Theta_4(g) = \Theta_4(S^n)$ if and only if (M,g) is conformally diffeomorphic to the standard sphere.*
(2) $\Theta_4(g)$ *is always achieved. Any maximizer f must be smooth and cannot change sign. If $f > 0$, then after scaling we have $G_P f = \frac{2}{n-4} f^{\frac{n-4}{n+4}}$ i.e. $Q_{f^{\frac{4}{n+4}}g} = 1$.*
(3) *If (M,g) is not conformally diffeomorphic to the standard sphere, then the set of all maximizers f for $\Theta_4(g)$, after normalizing with $\|f\|_{L^{\frac{2n}{n+4}}} = 1$, is compact in the C^∞ topology.*

The approach in Theorem 3.16 is motivated from the integral equations considered in [20, 21]. Integral equation formulation of the Q curvature equation had been used in [38].

At last, we note that compactness problem for constant Q curvature metrics in a fixed conformal class has been considered in [27, 32, 33, 37, 43].

4. Dimension 3

As we will see soon, the analysis of Q curvature equation in dimension 3 is very different from those in dimension greater than 4. On the other hand, we expect the scalar curvature and Q curvature plays more dominant role for the geometry of the conformal class and the topology of the underlying manifold in dimension 3 than in dimension greater than 4. Because of this, we will list problems in dimension 3 explicitly even though some of them are similar to those in Section 3.

In dimension 3, the Q curvature is given by

$$Q = -\frac{1}{4}\Delta R - 2\,|Rc|^2 + \frac{23}{32}R^2 \qquad (4.1)$$

$$= -\Delta J - 2\,|A|^2 + \frac{3}{2}J^2$$

$$= -\Delta J + 4\sigma_2(A) - \frac{1}{2}J^2,$$

here

$$J = \frac{R}{4}, \quad A = Rc - Jg. \qquad (4.2)$$

The Paneitz operator is given by

$$P\varphi = \Delta^2\varphi + 4div\left[Rc\left(\nabla\varphi, e_i\right)e_i\right] - \frac{5}{4}div\left(R\nabla\varphi\right) - \frac{1}{2}Q\varphi \qquad (4.3)$$

$$= \Delta^2\varphi + 4div\left(A\left(\nabla\varphi, e_i\right)e_i\right) - div\left(J\nabla\varphi\right) - \frac{1}{2}Q\varphi.$$

Here e_1, e_2, e_3 is a local orthonormal frame with respect to g. For any smooth positive function ρ,

$$P_{\rho^{-4}g}\varphi = \rho^7 P_g\left(\rho\varphi\right). \qquad (4.4)$$

Hence

$$Q_{\rho^{-4}g} = -2\rho^7 P_g\left(\rho\right). \qquad (4.5)$$

Problem 4.1: *Let (M, g) be a 3-dimensional smooth compact Riemannian manifold, can we always find a conformal metric $\tilde{g}$ such that $\tilde{Q}$ is either strictly positive, or identically zero, or strictly negative? Can we find a conformal invariant condition which is equivalent to the existence of a conformal metric with positive Q curvature? Same questions can be asked when "positive" is replaced by "negative" or "zero".*

Unfortunately, this simple-looking question only has a partial solution at this stage.

Theorem 4.2: *[25] Let (M, g) be a smooth compact 3-dimensional Riemannian manifold with $Y(g) > 0$, then the following statements are equivalent:*

(1) $\exists \tilde{g} \in [g]$ with $\tilde{Q} > 0$.
(2) $\ker P_g = 0$ and the Green's function $G_P(p, q) < 0$ for any $p, q \in M, p \neq q$.
(3) $\ker P_g = 0$ and there exists a $p \in M$ such that $G_P(p, q) < 0$ for $q \in M \setminus \{p\}$.

By transformation law (4.4) we know $\ker P_g = 0$ is a conformal invariant condition. Under this assumption, the Green's functions satisfy

$$G_{P, \rho^{-4}g}(p, q) = \rho(p)^{-1}\rho(q)^{-1}G_{P, g}(p, q). \qquad (4.6)$$

Hence the fact $G_P(p, q) < 0$ for $p \neq q$ is a conformal invariant condition. Theorem 4.2 is based on the following identity:

Theorem 4.3: *[25] Let (M, g) be a 3-dimensional smooth compact Riemannian manifold with $Y(g) > 0$, $p \in M$, then we have $G_{L,p}^{-1}\left|Rc_{G_{L,p}^4 g}\right|_g^2 \in$*

$L^1(M)$ *and*

$$P\left(G_{L,p}^{-1}\right) = -256\pi^2\delta_p + G_{L,p}^{-1}\left|Rc_{G_{L,p}^4 g}\right|_g^2 \qquad (4.7)$$

in distribution sense.

If $Y(g) > 0$, we write

$$H(p,q) = -\frac{G_L(p,q)^{-1}}{256\pi^2}, \qquad (4.8)$$

and

$$\Gamma_1(p,q) = \frac{G_L(p,q)^{-1}}{256\pi^2}\left|Rc_{G_{L,p}^4 g}\right|_g^2(q). \qquad (4.9)$$

Then (4.7) becomes

$$P_q H(p,q) = \delta_p(q) - \Gamma_1(p,q). \qquad (4.10)$$

Note that by the calculation in Section 2 of [25],

$$\Gamma_1(p,q) = O\left(\overline{pq}^{-1}\right), \qquad (4.11)$$

here $\overline{pq}$ denotes the distance between p and q.

If we let $\widetilde{g} = \rho^{-4}g$, here ρ is a positive smooth function, then for any smooth function φ on M,

$$T_{\widetilde{\Gamma}_1}(\varphi) = \rho^{-1}T_{\Gamma_1}(\rho\varphi). \qquad (4.12)$$

Hence $T_{\widetilde{\Gamma}_1}$ and T_{Γ_1} have the same spectrum and spectral radius.

Theorem 4.4: *Let (M,g) be a 3-dimensional smooth compact Riemannian manifold with $Y(g) > 0$, then*

$$\exists\widetilde{g} \in [g] \ \text{ with } \ \widetilde{Q} > 0. \Longleftrightarrow \text{ the spectral radius } r_\sigma(T_{\Gamma_1}) < 1.$$

Moreover, if $r_\sigma(T_{\Gamma_1}) < 1$, then $\ker P = 0$ and

$$G_P = H + \sum_{k=1}^{\infty}\Gamma_k * H, \qquad (4.13)$$

here

$$\Gamma_k = \Gamma_1 * \cdots * \Gamma_1 \ (k \text{ times}), \qquad (4.14)$$

H *and* Γ_1 *are given in (4.8) and (4.9). The convergence in (4.13) is uniform. In particular, $G_P \leq H$, moreover if $G_P(p,q) = H(p,q)$ for some p,q, then (M,g) is conformal equivalent to the standard S^3.*

Proof: The argument is basically same as the proof of Theorem 3.5. If there exists a $\widetilde{g} \in [g]$ with $\widetilde{Q} > 0$, by conformal invariance we can assume the background metric has positive Q curvature. By (4.10) for any smooth function φ,

$$\varphi = T_H \left(P\varphi \right) + T_{\Gamma_1} \left(\varphi \right). \tag{4.15}$$

Taking $\varphi = 1$ in (4.15) we get

$$\int_M \Gamma_1 \left(p, q \right) d\mu \left(q \right) = 1 + \frac{1}{2} \int_M H \left(p, q \right) Q \left(q \right) d\mu \left(q \right). \tag{4.16}$$

Hence for some α

$$\int_M \Gamma_1 \left(p, q \right) d\mu \left(q \right) \leq \alpha < 1$$

for all $p \in M$. It follows that

$$\| T_{\Gamma_1} \|_{\mathcal{L}(L^\infty, L^\infty)} \leq \alpha$$

and

$$r_\sigma \left(T_{\Gamma_1} \right) \leq \alpha < 1.$$

On the other hand, assume $r_\sigma \left(T_{\Gamma_1} \right) < \alpha < 1$, then we can find a constant k_0 such that for $k \geq k_0$,

$$\| T_{\Gamma_k} \|_{\mathcal{L}(L^\infty, L^\infty)} < \alpha^k.$$

It follows that

$$\int_M \Gamma_k \left(p, q \right) d\mu \left(q \right) < \alpha^k.$$

Using (4.11) we see for all $k \geq k_0 + 2$,

$$\| \Gamma_k \|_{L^\infty} \leq \alpha^{k-2} \| \Gamma_2 \|_{L^\infty} \leq c\alpha^k.$$

In particular, $\| \Gamma_k \|_{L^\infty} \to 0$ and

$$\| \Gamma_k * H \|_{L^\infty} \leq c \| \Gamma_k \|_{L^\infty} \leq c\alpha^k.$$

The remaining argument goes exactly the same as in the proof of Theorem 3.5. □

Remark 4.5: Indeed it follows from (4.16) that as long as $Y(g) > 0$ and

$$\int_M H \left(p, q \right) Q \left(q \right) d\mu \left(q \right) < 0$$

for all $p \in M$, then $r_\sigma \left(T_{\Gamma_1} \right) < 1$. In particular, this is the case when $Q \geq 0$ and not identically zero.

It is worth pointing out that if $\ker P = 0$, then because $\delta_p \in H^{-2}(M)$, we see $G_{P,p} \in H^2(M) \subset C^{\frac{1}{2}}(M)$, in particular the Green's function has a value at the pole, $G_{P,p}(p)$. This pole's value plays exactly the same role as the mass for classical Yamabe problem. If $Y(g) > 0$, $r_\sigma(T_{\Gamma_1}) < 1$ and (M, g) is not conformal diffeomorphic to the standard S^3, it follows from Theorem 4.4 that $G_P(p, q) < 0$ for all $p, q \in M$. On the other hand, on the standard S^3, the Green's function of Paneitz operator touches zero exactly at the pole and is negative away from the pole.

Problem 4.6: *Let (M, g) be a 3-dimensional smooth compact Riemannian manifold, can we find a metric $\widetilde{g} \in [g]$ such that $\widetilde{Q} = const$?*

Theorem 4.7: [24, 25] *Let (M, g) be a 3-dimensional smooth compact Riemannian manifold with $Y(g) > 0$ and $r_\sigma(T_{\Gamma_1}) < 1$, then there exists $\widetilde{g} \in [g]$ such that $\widetilde{Q} = 1$. Moreover, as long as (M, g) is not conformal diffeomorphic to the standard S^3, the set $\left\{ \widetilde{g} \in [g] : \widetilde{Q} = 1 \right\}$ is compact in C^∞ topology.*

Indeed let $\widetilde{g} = u^{-4}g$, then $\widetilde{Q} = 1$ becomes

$$Pu = -\frac{1}{2}u^{-7}, \quad u \in C^\infty(M), u > 0. \tag{4.17}$$

We can assume (M, g) is not conformal diffeomorphic to the standard S^3, then it follows from Theorem 4.4 that $\ker P = 0$ and $G_P(p, q) < 0$ for all $p, q \in M$. Let $K(p, q) = -G_P(p, q) > 0$, then (4.17) becomes

$$u = \frac{1}{2}T_K\left(u^{-7}\right). \tag{4.18}$$

For $0 \le t \le 1$, we consider a family of integral equations

$$u = \frac{1}{2}T_{(1-t)+tK}\left(u^{-7}\right). \tag{4.19}$$

Elementary *a priori* estimate for (4.19) based on the fact K is bounded and strictly positive together with a degree theory argument gives us Theorem 4.7 (see [24]). Note the proof of Theorem 4.7 is technically simpler than the proof of Theorem 3.15. This gives a partial solution to Problem 4.6.

To find more solutions to Problem 4.6, we turn our attention to variational methods. If we write $\widetilde{g} = \rho^{-4}g$, then the problem becomes

$$P\rho = const \cdot \rho^{-7}, \quad \rho \in C^\infty(M), \rho > 0. \tag{4.20}$$

For $u \in C^\infty (M)$, we denote

$$E (u, v) = \int_M Pu \cdot u d\mu \qquad (4.21)$$

$$= \int_M \left[(\Delta u)^2 - 4Rc (\nabla u, \nabla u) + \frac{5}{4} R |\nabla u|^2 - \frac{1}{2} Q u^2 \right] d\mu$$

$$= \int_M \left[(\Delta u)^2 - 4A (\nabla u, \nabla u) + J |\nabla u|^2 - \frac{1}{2} Q u^2 \right] d\mu.$$

It is clear that $E (u)$ extends continuously to $u \in H^2 (M)$. Sobolev embedding theorem tells us $H^2 (M) \subset C^{\frac{1}{2}} (M)$, hence we can set

$$Y_4 (g) = \inf_{u \in H^2 (M), u > 0} E (u) \left\| u^{-1} \right\|_{L^6}^2 = -\frac{1}{2} \sup_{\widetilde{g} \in [g]} \widetilde{\mu} (M)^{\frac{1}{3}} \int_M \widetilde{Q} d\widetilde{\mu}. \quad (4.22)$$

$Y_4 (g)$ is a conformal invariant similar to $Y (g)$. But unlike $Y (g)$, it is not clear anymore whether $Y_4 (g)$ is finite or not.

Problem 4.8: *Let (M, g) be a 3-dimensional smooth compact Riemannian manifold, do we have $Y_4 (g) > -\infty$? Is $Y_4 (g)$ always achieved?*

To better understand the problem, following [22], we start with some basic analysis. Let u_i be a minimizing sequence for (4.22). By scaling, we can assume $\|u_i\|_{L^2} = 1$. By Holder inequality, we have

$$c = \|1\|_{L^{\frac{3}{2}}} \le \|u_i\|_{L^2} \left\| u_i^{-1} \right\|_{L^6},$$

hence

$$\left\| u_i^{-1} \right\|_{L^6} \ge c > 0.$$

It follows that $E (u_i) \le c$ and hence $\|u_i\|_{H^2} \le c$. After passing to a subsequence, we can find $u \in H^2 (M)$ such that $u_i \rightharpoonup u$ weakly in $H^2 (M)$. It follows that $\|u\|_{L^2} = 1$ and $u \ge 0$.

If $u > 0$, then by lower semicontinuity we know u is a minimizer. On the other hand if u touches zero somewhere, then

$$\infty = \left\| u^{-1} \right\|_{L^6} \le \lim_{i \to \infty} \inf \left\| u_i^{-1} \right\|_{L^6},$$

hence

$$E (u) \le \lim_{i \to \infty} \inf E (u_i) \le 0.$$

If we can rule out the second case, then $Y_4 (g)$ is achieved.

Definition 4.9: [22] Let (M, g) be a 3-dimensional smooth compact Riemannian manifold. If $u \in H^2 (M)$ with $u \ge 0$ and $u = 0$ somewhere would imply $E(u) \geq 0$, then we say the metric g (or the associated Paneitz operator) satisfies condition NN^+. If $u \in H^2(M)$ is a nonzero function with $u \geq 0$ and $u = 0$ somewhere would imply $E(u) > 0$, then we say the metric g satisfies condition P^+.

Theorem 4.10: [22] *Let (M, g) be a 3-dimensional smooth compact Riemannian manifold. Then we have*

$$Y_4(g) \text{ is finite} \Rightarrow g \text{ satisfies } NN^+$$

and

$$g \text{ satisfies } P^+ \Rightarrow Y_4(g) \text{ is achieved and hence finite.}$$

Note condition P^+ is clearly satisfied when $P > 0$. In this case, Theorem 4.10 was proved in [46]. Here is an example when we have positivity of the Paneitz operator.

Lemma 4.11: [22] *If $Y(g) > 0, \sigma_2(A) > 0$, $Q \leq 0$ and not identically zero, then $P > 0$.*

Examples satisfying assumptions in Lemma 4.11 can be found in Berger spheres (see [22]). Here we give another criterion for positivity in the same spirit as Theorem 1.6 of [9].

Lemma 4.12: *If $\sigma_2(A) < 0$ and $2Jg \geq A$ (note this implies $J \geq 0$), then $P > 0$.*

Proof: Let

$$\Theta = D^2 u - \frac{\Delta u}{3} g$$

be the traceless Hessian and

$$\mathring{A} = A - \frac{J}{3} g$$

be the traceless Schouten tensor. For convenience, we use $A \sim B$ to mean $\int_M A d\mu = \int_M B d\mu$. First, we derive the Bochner identity,

$$(\Delta u)^2 = u_{ii} u_{jj} \sim -u_{iij} u_j = -(u_{iji} - R_{ijik} u_k) u_j \sim u_{ij} u_{ij} + Rc_{jk} u_j u_k$$

$$= |D^2 u|^2 + Rc(\nabla u, \nabla u) = |D^2 u|^2 + A(\nabla u, \nabla u) + J |\nabla u|^2.$$

Hence

$$(\Delta u)^2 \sim |\Theta|^2 + \frac{(\Delta u)^2}{3} + A(\nabla u, \nabla u) + J |\nabla u|^2.$$

In another way,

$$(\Delta u)^2 \sim \frac{3}{2}|\Theta|^2 + \frac{3}{2}A\left(\nabla u, \nabla u\right) + \frac{3}{2}J|\nabla u|^2.$$

The next step is to remove the ΔJ term in Q curvature. Note that

$$-\Delta J \cdot u^2 = -J_{ii}u^2 \sim 2J_i u \cdot u_i = 2A_{ijj}u \cdot u_i \sim -2A_{ij}u_i u_j - 2A_{ij}u_{ij}u$$

$$= -2A_{ij}u_i u_j - 2A_{ij}\Theta_{ij}u - \frac{2}{3}Ju\Delta u$$

$$= -2A_{ij}u_i u_j - 2A_{ij}\Theta_{ij}u - \frac{1}{3}J\left(\Delta u^2 - 2|\nabla u|^2\right)$$

$$\sim -2A_{ij}u_i u_j - 2A_{ij}\Theta_{ij}u - \frac{1}{3}\Delta J \cdot u^2 + \frac{2}{3}J|\nabla u|^2.$$

Hence

$$-\Delta J \cdot u^2 \sim -3A\left(\nabla u, \nabla u\right) - 3A_{ij}\Theta_{ij}u + J|\nabla u|^2$$

$$= -3A\left(\nabla u, \nabla u\right) - 3\mathring{A}_{ij}\Theta_{ij}u + J|\nabla u|^2.$$

It follows that

$$(\Delta u)^2 + J|\nabla u|^2 - 4A\left(\nabla u, \nabla u\right) - \frac{1}{2}Qu^2$$

$$\sim \frac{3}{2}|\Theta|^2 + \frac{3}{2}\mathring{A}_{ij}\Theta_{ij}u + 2J|\nabla u|^2 - A\left(\nabla u, \nabla u\right) - \frac{3}{4}J^2u^2 + |A|^2 u^2$$

$$= \frac{3}{2}\left|\Theta + \frac{1}{2}u\mathring{A}\right|^2 + 2J|\nabla u|^2 - A\left(\nabla u, \nabla u\right) - \frac{5}{8}\left(J^2 - |A|^2\right)u^2.$$

In other words

$$E\left(u\right) = \frac{3}{2}\int_M \left|\Theta + \frac{1}{2}u\mathring{A}\right|^2 d\mu + \int_M \left[2J|\nabla u|^2 - A\left(\nabla u, \nabla u\right)\right]d\mu$$

$$- \frac{5}{8}\int_M \left(J^2 - |A|^2\right)u^2 d\mu.$$

The positivity follows. $\square$

The assumption in Lemma 4.12 is satisfied by $S^2 \times S^1$ with the product metric and some Berger's spheres (see [22]).

Conditions P$^+$ and NN$^+$ are hard to check in general, on the other hand, they are hard to use too. The closely related conditions P and NN can be introduced.

Definition 4.13: [22] Let (M, g) be a 3-dimensional smooth compact Riemannian manifold. If $u \in H^2(M)$ with $u = 0$ somewhere would imply $E(u) \geq 0$, then we say the metric g (or the associated Paneitz operator) satisfies condition NN. If $u \in H^2(M)$ is a nonzero function with $u = 0$ somewhere would imply $E(u) > 0$, then we say the metric g satisfies condition P.

Condition NN can be used to identify the limit function u, when u touches zero in the brief discussion after Problem 4.8 (see [22]).

The standard sphere S^3 does not satisfy condition P$^+$. Indeed, let x be the coordinate given by the stereographic projection with respect to north pole N, then the Green's function of P at N can be written as

$$G_N = -\frac{1}{4\pi} \frac{1}{\sqrt{|x|^2 + 1}}. \tag{4.23}$$

In particular, $E(G_N) = G_N(N) = 0$.

Theorem 4.14: [48] $Y_4(S^3, g_{S^3})$ *is achieved at the standard metric.*

Indeed [48] shows $Y_4(S^3)$ is achieved by the method of symmetrization. All the critical points are classified by [44]. In [19, 22], several different approaches are given. The main ingredient is the following observation:

Lemma 4.15: [22] *Let $N \in S^3$ be the north pole, $u \in H^2(S^3)$ such that $u(N) = 0$. Denote x as the coordinate given by the stereographic projection with respect to N and*

$$\tau = \sqrt{\frac{|x|^2 + 1}{2}}.$$

Then we know $\Delta(\tau u) \in L^2(\mathbb{R}^3)$ and

$$E(u) = \int_{\mathbb{R}^3} |\Delta(\tau u)|^2 \, dx, \tag{4.24}$$

here Δ is the Euclidean Laplacian.

In particular, S^3 satisfies NN. The only functions touching 0 and having nonpositive energy are constant multiples of Green's functions.

To help understanding the condition NN, in [23], new quantities $\nu(M, g, p)$ and $\nu(M, g)$ are introduced. Let (M, g) be a 3-dimensional smooth compact Riemannian manifold, for any $p \in M$, define

$$\nu(M, g, p) = \inf \left\{ \frac{E(u)}{\int_M u^2 d\mu} : u \in H^2(M) \setminus \{0\}, u(p) = 0 \right\}. \tag{4.25}$$

When no confusion could arise we denote it as $\nu\left(g,p\right)$ or ν_p. We also define

$$\nu\left(M,g\right) \tag{4.26}$$
$$= \inf_{p\in M} \nu\left(M,g,p\right)$$
$$= \inf\left\{\frac{E\left(u\right)}{\int_M u^2 d\mu} : u \in H^2\left(M\right)\setminus\{0\}, u\left(p\right) = 0 \text{ for some } p\right\}.$$

The importance of $\nu\left(M,g\right)$ lies in that g satisfies condition P if and only if $\nu\left(g\right) > 0$ and it satisfies condition NN if and only if $\nu\left(g\right) \geq 0$. It follows from Lemma 4.15 that $\nu\left(S^3, g_{S^3}\right) = 0$. A closely related fact is that the Green's function of Paneitz operator on S^3 vanishes at the pole. In [23], first and second variations of $G_P\left(N,N\right)$ and $\nu\left(S^3, g, N\right)$ are calculated.

Theorem 4.16: [23] *Let g be the standard metric on S^3 and h be a smooth symmetric $(0,2)$ tensor. Denote $x = \pi_N$, the stereographic projection with respect to N and*

$$\tau = \sqrt{\frac{|x|^2 + 1}{2}}.$$

Let G_{g+th} be the Green's function of the Paneitz operator P_{g+th}, then

$$\left.\partial_t\right|_{t=0} G_{g+th}\left(N,N\right) = 0 \tag{4.27}$$

and

$$\left.\partial_t^2\right|_{t=0} G_{g+th}\left(N,N\right) \tag{4.28}$$
$$= -\frac{1}{64\pi^2}\int_{\mathbb{R}^3}\left(\sum_{ij}\left(\theta_{ikjk} + \theta_{jkik} - (tr\theta)_{ij} - \Delta\theta_{ij}\right)^2 - \frac{3}{2}\left(\theta_{ijij} - \Delta tr\theta\right)^2\right)dx.$$

Here $\theta = \tau^4 h$ and the derivatives θ_{ikjk} etc are partial derivatives in $\mathbb{R}^3$.
    In particular,

$$\left.\partial_t^2\right|_{t=0} G_{g+th}\left(N,N\right) \leq 0. \tag{4.29}$$

Moreover, $\left.\partial_t^2\right|_{t=0} G_{g+th}\left(N,N\right) = 0$ if and only if $h = L_X g + f \cdot g$ for some smooth vector fields X and smooth function f on S^3.
    For $\nu\left(g + th, N\right)$, we have

$$\left.\partial_t\right|_{t=0} \nu\left(g + th, N\right) = 0 \tag{4.30}$$

and

$$\left.\partial_t^2\right|_{t=0} \nu\left(g + th, N\right) = -16\left.\partial_t^2\right|_{t=0} G_{g+th}\left(N,N\right). \tag{4.31}$$

In [23], a close relation between condition NN and the second eigenvalue of Paneitz operator is given.

Theorem 4.17: [23] *Let (M, g) be a 3-dimensional smooth compact Riemannian manifold with $Y(g) > 0$ and $r_\sigma(T_{\Gamma_1}) < 1$, then the following statements are equivalent:*

(1) $Y_4(g) > -\infty$.
(2) $\lambda_2(P) > 0$.
(3) $\nu(g) \geq 0$ i.e. (M, g) satisfies condition NN.

For condition P, there is a similar statement.

Corollary 4.18: [23] *Let (M, g) be a 3-dimensional smooth compact Riemannian manifold with $Y(g) > 0$ and $r_\sigma(T_{\Gamma_1}) < 1$. If (M, g) is not conformal diffeomorphic to the standard S^3, then the following statements are equivalent:*

(1) $Y_4(g) > -\infty$.
(2) $\lambda_2(P) > 0$.
(3) $\nu(g) > 0$ i.e. (M, g) satisfies condition P.

These statements make the finiteness of $Y_4(g)$ and condition NN more meaningful.

Problem 4.19: *Let (M, g) be a 3-dimensional smooth compact Riemannian manifold, does g always satisfy condition NN? Does metric with positive Yamabe invariant always satisfy condition NN?*

This seems to be a difficult question. We only have a partial answer.

Theorem 4.20: [24, 25] *Assume M is a smooth compact 3-dimensional manifold, denote*

$$\mathcal{M} = \{g : Y(g) > 0, \, r_\sigma(T_{\Gamma_1}) < 1\}, \tag{4.32}$$

endowed with C^∞ topology. Let $\mathcal{N}$ be a path connected component of $\mathcal{M}$. If there is a metric in $\mathcal{N}$ satisfying condition NN, then every metric in $\mathcal{N}$ satisfies condition NN. Hence as long as the metric is not conformal equivalent to the standard S^3, it satisfies condition P.

Here we describe an application of above discussions. Let M be a 3-dimensional smooth compact manifold, $\beta \in \mathbb{R}$, we define a functional

$$F_\beta(g) = \int_M Q d\mu + \beta \int_M J^2 d\mu. \tag{4.33}$$

Calculation shows that the critical metric of F_β restricted to a fixed conformal class with unit volume constraint is given by

$$Q + 2\beta \Delta J + \beta J^2 = const. \tag{4.34}$$

Proposition 4.21: *Assume (M, g) is a 3-dimensional smooth compact Riemannian manifold satisfying condition P^+, $\beta \leq 0$, then*

$$\sup_{\widetilde{g} \in [g]} \widetilde{\mu}(M)^{\frac{1}{3}} F_\beta(\widetilde{g}) \tag{4.35}$$

is achieved.

Proof: For any positive smooth function u,

$$F_\beta(u^{-4}g) = -2 \int_M Pu \cdot u d\mu + \beta \int_M u^4 \left[-2\Delta(u^{-1}) + Ju^{-1} \right]^2 d\mu. \tag{4.36}$$

Define

$$\Phi_\beta(u) = \int_M Pu \cdot u d\mu - \frac{\beta}{2} \int_M u^4 \left[-2\Delta(u^{-1}) + Ju^{-1} \right]^2 d\mu, \tag{4.37}$$

then

$$\sup_{\widetilde{g} \in [g]} \widetilde{\mu}(M)^{\frac{1}{3}} F_\beta(\widetilde{g}) = -2 \inf_{\substack{u \in C^\infty(M) \\ u > 0}} \left\| u^{-1} \right\|_{L^6}^2 \Phi_\beta(u). \tag{4.38}$$

Let

$$m = \inf_{\substack{u \in H^2(M) \\ u > 0}} \left\| u^{-1} \right\|_{L^6}^2 \Phi_\beta(u). \tag{4.39}$$

We claim m is achieved. Indeed

$$\Phi_\beta(u) = \int_M \left[(\Delta u)^2 - 4A(\nabla u, \nabla u) + J |\nabla u|^2 - \frac{1}{2} Q u^2 \right] d\mu \tag{4.40}$$

$$- 2\beta \int_M \left(\Delta u - 2u^{-1} |\nabla u|^2 + \frac{J}{2} u \right)^2 d\mu.$$

Assume $u_i \in H^2(M)$, $u_i > 0$ is a minimizing sequence, by scaling we can assume $\max_M u_i = 1$. Then

$$\left\| u_i^{-1} \right\|_{L^6}^2 \Phi_\beta(u_i) \to m.$$

It follows from $\beta \leq 0$ that

$$\left\| u_i^{-1} \right\|_{L^6}^2 E(u_i) \leq c.$$

Hence $E(u_i) \leq c$. Together with the fact $0 < u_i \leq 1$ we get $\|u_i\|_{H^2(M)} \leq c$. After passing to a subsequence, we can assume $u_i \rightharpoonup u$ weakly in $H^2(M)$. Then $u_i \to u$ uniformly. It follows that $\max_M u = 1$ and $u \geq 0$. We claim u cannot touch 0. Indeed if u touches zero somewhere, then since $u \in H^2(M)$ we see $\int_M u^{-6} d\mu = \infty$. It follows from Fatou's lemma that

$$\lim_{i\to\infty} \inf \int_M u_i^{-6} d\mu \geq \int_M u^{-6} d\mu = \infty.$$

Hence

$$\lim_{i\to\infty} \sup E(u_i) \leq 0.$$

It follows that $E(u) \leq 0$, this contradicts with condition P^+. The fact $u > 0$ follows. To continue, we observe that $u_i \to u$ in $W^{1,p}(M)$ for $p < 6$. Hence

$$\Phi_\beta(u) \leq \lim_{i\to\infty} \inf \Phi_\beta(u_i).$$

It follows that

$$\left\|u^{-1}\right\|_{L^6}^2 \Phi_\beta(u) \leq \lim_{i\to\infty} \inf \left\|u_i^{-1}\right\|_{L^6}^2 \Phi_\beta(u_i) = m.$$

u is a minimizer. Calculation shows for any $\varphi \in C^\infty(M)$,

$$\int_M uP\varphi d\mu - 2\beta \int_M \left(\Delta u - 2u^{-1}|\nabla u|^2 + \frac{J}{2}u\right)$$
$$\cdot \left(\Delta \varphi - 4u^{-1}\langle \nabla u, \nabla \varphi\rangle + 2u^{-2}|\nabla u|^2 \varphi + \frac{J}{2}\varphi\right) d\mu$$
$$= const \int_M u^{-7}\varphi d\mu.$$

Standard bootstrap method shows $u \in C^\infty(M)$. Proposition 4.21 follows. $\square$

Corollary 4.22: *Let (M, g) be a 3-dimensional smooth compact Riemannian manifold satisfying condition P^+, $Y(g) > 0$ and*

$$\int_M Qd\mu - \frac{1}{6}\int_M J^2 d\mu \geq 0, \tag{4.41}$$

then the universal cover of M is diffeomorphic to S^3.

Remark 4.23: (4.41) is the same as

$$\int_M |E|^2 d\mu \leq \frac{1}{48}\int_M R^2 d\mu, \tag{4.42}$$

here $E = Rc - \frac{R}{3}g$ is the traceless Ricci tensor.

Proof: It follows from Proposition 4.21 that

$$\kappa = \sup_{\tilde{g} \in [g]} \tilde{\mu}(M)^{\frac{1}{3}} \left(\int_M \tilde{Q} d\tilde{\mu} - \frac{1}{6} \int_M \tilde{J}^2 d\tilde{\mu} \right) \tag{4.43}$$

is achieved. By (4.41), we know $\kappa \geq 0$. Without loss of generality, we can assume the background metric g is a maximizer and it has volume 1. Then

$$Q - \frac{1}{3} \Delta J - \frac{1}{6} J^2 = const. \tag{4.44}$$

Integrating both sides we get

$$Q - \frac{1}{3} \Delta J - \frac{1}{6} J^2 = \kappa \geq 0. \tag{4.45}$$

In another way it is

$$-\frac{4}{3} \Delta J - 2 \left| \mathring{A} \right|^2 + \frac{2}{3} J^2 = \kappa. \tag{4.46}$$

Here $\mathring{A} = A - \frac{J}{3} g$ is the traceless Schouten tensor. Since the conformal Laplacian is given by $L = -8\Delta + 4J$,

$$LJ = 6\kappa + 12 \left| \mathring{A} \right|^2 \geq 0. \tag{4.47}$$

Since $Y(g) > 0$, we see the Green's function of L must be positive, hence either $J > 0$ or $J \equiv 0$. The latter case contradicts with the fact $Y(g) > 0$. So the scalar curvature must be strictly positive. Finally

$$\int_M \sigma_2(A) \, d\mu = \frac{1}{4} \int_M \left(Q + \frac{1}{2} J^2 \right) d\mu = \frac{\kappa}{4} + \frac{1}{6} \int_M J^2 d\mu > 0. \tag{4.48}$$

It follows from a result in [6, 14] that the universal cover of M is diffeomorphic to S^3. $\qquad \square$

The above example also shows the interest in the following question:

Problem 4.24: *Let (M, g) be a 3-dimensional smooth compact Riemannian manifold, can we find a conformal invariant condition which is equivalent to the existence of a conformal metric with positive scalar and Q curvature?*

We remark that by modifying the technique in [18], it is shown in [24] that on 3-dimensional smooth compact Riemannian manifolds (M, g) with $R > 0$ and $Q > 0$, if $\tilde{g} \in [g]$ satisfies $\tilde{Q} > 0$, then $\tilde{R} > 0$ too.

Acknowledgments

We would like to thank Jie Qing and Xingwang Xu for their invitations to talk about the results in this set of lecture notes.

The research of Yang is supported by NSF grant 1104536.

References

1. D. Adams. *A sharp inequality of J. Moser for higher order derivatives.* Ann. of Math. (2) **128** (1988), no. 2, 385–398.
2. T. Aubin. *Fonction de Green et valeurs propres du laplacien.* J. Math. Pures Appl. (9) **53** (1974), 347–371.
3. T. Aubin. Nonlinear analysis on manifolds, Monge-Ampere equations. Springer Verlag, New York, 1982.
4. T. Branson. *Differential operators canonically associated to a conformal structure.* Math. Scand. **57** (1985), no. 2, 293–345.
5. T. Branson, S. Y. Chang and P. C. Yang. *Estimates and extremals for zeta function determinants on four-manifolds.* Comm. Math. Phys. **149** (1992), no. 2, 241–262.
6. G. Catino and Z. Djadli. *Conformal deformations of integral pinched 3-manifolds.* Adv. Math. **223** (2010), no. 2, 393–404.
7. S. Y. Chang, M. J. Gursky and P. C. Yang. *An equation of Monge-Ampere type in conformal geometry, and four-manifolds of positive Ricci curvature.* Ann. of Math. (2) **155** (2002), no. 3, 709–787.
8. S. Y. Chang, M. J. Gursky and P. C. Yang. *A conformally invariant sphere theorem in four dimensions.* Publ. Math. Inst. Hautes Etudes Sci. No. **98** (2003), 105–143.
9. S. Y. Chang, F. B. Hang and P. C. Yang. *On a class of locally conformally flat manifolds.* Int. Math. Res. Not. **2004**, no. 4, 185–209.
10. S. Y. Chang and P. C. Yang. *Extremal metrics of zeta function determinants on 4-manifolds.* Ann. of Math. (2) **142** (1995), no. 1, 171–212.
11. W. X. Chen, C. M. Li and B. Ou. *Classification of solutions for an integral equation.* Comm Pure Appl Math **59** (2006), no. 3, 330–343.
12. Z. Djadli and A. Malchiodi. *Existence of conformal metrics with constant Q curvature.* Ann. of Math. (2) **168** (2008), no. 3, 813–858.
13. L. Fontana. *Sharp borderline Sobolev inequalities on compact Riemannian manifolds.* Comment. Math. Helv. **68** (1993), no. 3, 415–454.
14. Y. X. Ge, C. S. Lin and G. F. Wang. *On the σ_2-scalar curvature.* J. Differential Geom. **84** (2010), no. 1, 45–86.
15. M. J. Gursky. *The Weyl functional, de Rham cohomology, and Kahler-Einstein metrics.* Ann. of Math. (2) **148** (1998), no. 1, 315–337.
16. M. J. Gursky. *The principal eigenvalue of a conformally invariant differential operator.* Comm. Math. Phys. **207** (1999), no. 1, 131–143.
17. M. J. Gursky, F. B. Hang and Y. J. Lin. *Riemannian manifolds with positive Yamabe invariant and Paneitz operator.* International Mathematics Research Notices 2015; doi:10.1093/imrn/rnv176.

18. M. J. Gursky and A. Malchiodi. *A strong maximum principle for the Paneitz operator and a nonlocal flow for the Q curvature.* J. Eur. Math. Soc. **17** (2015), 2137–2173.

19. F. B. Hang. *On the higher order conformal covariant operators on the sphere.* Commun Contemp Math. **9** (2007), no. 3, 279–299.

20. F. B. Hang, X. D. Wang and X. D. Yan. *Sharp integral inequalities for harmonic functions.* Comm. Pure Appl. Math. **61** (2008), no. 1, 54–95.

21. F. B. Hang, X. D. Wang and X. D. Yan. *An integral equation in conformal geometry.* Ann. Inst. H. Poincare Anal. Non Lineaire **26** (2009), no. 1, 1–21.

22. F. B. Hang and P. C. Yang. *The Sobolev inequality for Paneitz operator on three manifolds.* Calculus of Variations and PDE. **21** (2004), 57–83.

23. F. B. Hang and P. C. Yang. Paneitz operator for metrics near S^3. Preprint (2014), arXiv:1504.02032.

24. F. B. Hang and P. C. Yang. *Q curvature on a class of 3 manifolds.* Comm Pure Appl Math, to appear, DOI: 10.1002/cpa.21559.

25. F. B. Hang and P. C. Yang. *Sign of Green's function of Paneitz operators and the Q curvature.* International Mathematics Research Notices. IMRN **2015**, no. 19, 9775–9791.

26. F. B. Hang and P. C. Yang. Q curvature on a class of manifolds with dimension at least 5. Comm Pure Appl Math, to appear, DOI: 10.1002/cpa.21623.

27. E. Hebey and F. Robert. *Compactness and global estimates for the geometric Paneitz equation in high dimensions.* Electron Res Ann Amer Math Soc. **10** (2004), 135–141.

28. E. Hebey and F. Robert. *Asymptotic analysis for fourth order Paneitz equations with critical growth.* Adv Calc Var. **4** (2011), no. 3, 229–275.

29. E. Humbert and S. Raulot. *Positive mass theorem for the Paneitz-Branson operator.* Calculus of Variations and PDE. **36** (2009), 525–531.

30. J. L. Kazdan and F. W. Warner. *Scalar curvature and conformal deformation of Riemannian structure.* J. Diff. Geom. **10** (1975), 113–134.

31. J. M. Lee and T. H. Parker. *The Yamabe problem.* Bull AMS. **17** (1987), no. 1, 37–91.

32. G. Li. A compactness theorem on Branson's Q-curvature equation. Preprint (2015), arXiv:1505.07692.

33. Y. Y. Li and J. G. Xiong. Compactness of conformal metrics with constant Q-curvature. I. Preprint (2015), arXiv:1506.00739.

34. J. Moser. *A sharp form of an inequality by N. Trudinger.* Indiana Univ. Math. J. **20** (1970/71), 1077–1092.

35. R. O'Neil. *Convolution operators and L(p,q) spaces.* Duke Math. J. **30** (1963), 129–142.

36. S. Paneitz. A quartic conformally covariant differential operator for arbitrary pseudo-Riemannian manifolds. Preprint (1983), arXiv:0803.4331.

37. J. Qing and D. Raske. *Compactness for conformal metrics with constant Q curvature on locally conformally flat manifolds.* Calc. Var. Partial Differential Equations **26** (2006), no. 3, 343–356.

38. J. Qing and D. Raske. *On positive solutions to semilinear conformally invariant equations on locally conformally flat manifolds.* Int Math Res Not. Art.

id 94172 (2006).

39. F. Robert. Fourth order equations with critical growth in Riemannian geometry. Unpublished notes. Available at http://www.iecn.u-nancy.fr/~frobert/LectRobertFourth.pdf.

40. R. Schoen. *Conformal deformation of a Riemannian metric to constant scalar curvature.* J. Differential Geom. **20** (1984), no. 2, 479–495.

41. R. Schoen and S. T. Yau. *Conformally flat manifolds, Kleinian groups and scalar curvature.* Inventiones Mathematicae **92** (1988) no. 1, 47–71.

42. N. Trudinger. *Remarks concerning the conformal deformation of Riemannian structures on compact manifolds.* Ann. Scuola Norm. Sup. Pisa (3) **22** (1968), 265–274.

43. J. C. Wei and C. Y. Zhao. *Non-compactness of the prescribed Q-curvature problem in large dimensions.* Calc. Var. Partial Differential Equations **46** (2013), no. 1-2, 123–164.

44. X. W. Xu. *Exact solutions of nonlinear conformally invariant integral equations in $\mathbb{R}^3$.* Adv. Math. **194** (2005), no. 2, 485–503.

45. X. W. Xu and P. C. Yang. *Positivity of Paneitz operators.* Discrete Contin. Dynam. Systems **7** (2001), no. 2, 329–342.

46. X. W. Xu and P. C. Yang. *On a fourth order equation in 3-D.* ESAIM: Control, optimization and Calculus of Variations **8** (2002) 1029–1042.

47. H. Yamabe. *On a deformation of Riemannian structures on compact manifolds.* Osaka Math. J. **12** (1960), 21–37.

48. P. C. Yang and M. J. Zhu. *Sharp inequality for Paneitz operator on S^3.* ESAIM: Control, Optim. and Calc. of Var. **10** (2004) 211–223.

AN INTRODUCTION TO THE FINITE AND INFINITE DIMENSIONAL REDUCTION METHODS

Manuel del Pino

Departamento de Ingeniería Matemática and CMM
Universidad de Chile, Casilla 170 Correo 3
Santiago, Chile
delpino@dim.uchile.cl

Juncheng Wei

Department of Mathematics
University of British Columbia
Vancouver, BC V6T 1Z2, Canada
jcwei@math.ubc.ca

We give an introductory description of the two gluing methods: finite dimensional and infinite dimensional. In each case, we use a model problem to illustrate the ideas.

1. Part I: Finite-Dimensional Reduction Method

1.1. *Introduction: What is finite dimensional Liapunov-Schmidt reduction method?*

We briefly introduce the abstract set-up of the finite dimensional Lyapunov-Schmidt reduction (although it is always used in a framework that occurs often in bifurcation theory).

Let X, Y be Banach spaces and $S(u)$ be a C^1 nonlinear map from X to Y. To find a solution to the nonlinear equation

$$S(u) = 0, \tag{1.1}$$

a natural way is to find approximations first and then to look for genuine solutions as (small) perturbations of approximations. Assume that U_λ are the approximations, where $\lambda \in \Lambda$ is the parameter (we think of Λ as the configuration space). Writing $u = U_\lambda + \phi$, then solving $S(u) = 0$ amounts to solving

$$L[\phi] + E + N(\phi) = 0, \tag{1.2}$$

where

$$L[\phi] = S'(U_\lambda)[\phi], \; E = S(U_\lambda), \; \text{and} \; N(\phi) = S(U_\lambda + \phi) - S(U_\lambda) - S'(U_\lambda)[\phi].$$

Here $S'(U_\lambda)$ stands for the Fréchet derivative of S at U_λ, E denotes the error of approximation, and $N(\phi)$ denotes the nonlinear term. In order to solve (1.2), we try to invert the linear operator L so that we can rephrase the problem as a fixed point problem. That is, when L has a uniformly bounded inverse in a suitable space, one can rewrite the equation (1.2) as

$$\phi = -L^{-1}[E + N(\phi)] = \mathcal{A}(\phi).$$

What is left is to use fixed point theorems such as contraction mapping theorem.

The finite dimensional Lyapunov-Schmidt reduction deals with the situation when the linear operator L is Fredholm and its eigenfunction space associated to small eigenvalues has finite dimensional. Assuming that $\{\mathcal{Z}_1, \ldots, \mathcal{Z}_n\}$ is a basis of the eigenfunction space associated to small eigenvalues of L, we can divide the procedure of solving (1.2) into two steps: [(i)] solving the projected problem for any $\lambda \in \Lambda$,

$$\begin{cases} L[\phi] + E + N(\phi) = \sum\limits_{j=1}^{n} c_j \mathcal{Z}_j, \\ \langle \phi, \mathcal{Z}_j \rangle = 0, \; \forall\, j = 1, \ldots, n, \end{cases}$$

where c_j may be constant or function depending on the form of $\langle \phi, \mathcal{Z}_j \rangle$.

[(ii)] solving the reduced problem

$$c_j(\lambda) = 0, \; \forall\, j = 1, \ldots, n,$$

by adjusting λ in the configuration space.

The original finite dimensional Liapunov-Schmidt reduction method was first introduced in a seminal paper by Floer and Weinstein [27] in their construction of single bump solutions to one-dimensional nonlinear Schrodinger equations (Oh [54] generalized to high dimensional case)

$$\epsilon^2 \Delta u - V(x)u + u^p = 0, u > 0, \; u \in H^1(\mathbb{R}^N). \tag{1.3}$$

On the other hand, Bahri [3] and Bahri-Coron [4] developed the reduction method for critical exponent problems. In the last fifteen years, there are renewed efforts in refining the finite dimensional reduction method by many authors. When combined with variational methods, this reduction becomes "localized energy method". For subcritical exponent problems, we refer to Ambrosetti-Malchiodi [1], Gui-Wei [28], Malchiodi [48], Li-Nirenberg [41], Lin-Ni-Wei [42], Ao-Wei-Zeng [2], Wei-Yan [63] and the references therein. The localized energy method in degenerate setting is done by Byeon-Tanaka [6, 7]. For critical exponents, we refer to Bahri-Li-Rey [5], Del Pino-Felmer-Musso [17], Del Pino-Kowalczyk-Musso [18], Li-Wei-Xu [40], Rey-Wei [56, 57] and Wei-Yan [64] and the references therein. Many new features of the finite dimensional reduction are found in the references mentioned.

In the following, we shall use the model problem (1.3) to give an introductory description of this method.

1.2. *Model problem: Schrodinger equation in dimension N*

We start with the following model problem to illustrate the idea of finite dimensional reduction method:

$$\begin{cases} \varepsilon^2 \Delta u - V(x)u + u^p = 0 & \text{in } \mathbb{R}^N \\ 0 < u \text{ in } \mathbb{R}^N, \quad u(x) \to 0, \quad \text{as } |x| \to \infty. \end{cases} \quad (1.4)$$

The basic assumption on the exponent is that $1 < p < \infty$ if $N \le 2$, and $1 < p < \frac{N+2}{N-2}$ if $N \ge 3$. (More general nonlinearity can be dealt with similarly.) Without loss of generality, we assume that the function $V(x)$ is a positive function satisfying

$$0 < \alpha \le V(x) \le \beta < +\infty. \quad (1.5)$$

The basic building block that we consider is

$$\begin{cases} \Delta w - w + w^p = 0 & \text{in } \mathbb{R}^N \\ 0 < w \text{ in } \mathbb{R}^N, \quad w(x) \to 0, \quad \text{as } |x| \to \infty. \end{cases} \quad (1.6)$$

We look for a solution $w = w(|x|)$, a radially symmetric solution. $w(r)$ satisfies the ordinary differential equation

$$\begin{cases} w'' + \frac{N-1}{r}w' - w + w^p = 0 & r \in (0, \infty) \\ w'(0) = 0, 0 < w \text{ in } (0, \infty) \ w(|x|) \to 0, \quad \text{as } |x| \to \infty. \end{cases} \quad (1.7)$$

We collect the following basic properties of w, whose proof can be found in the appendix of the book [62].

Proposition 1.1: *(a) There exist a solution $w(r)$ to (1.7);*
(b) $w(r)$ satisfies the decay estimate $w(r) = A_0 r^{-\frac{N-1}{2}} e^r (1 + O(\frac{1}{r}))$;
(c) $w(r)$ is nondegenerate, i.e., the only bounded solution to

$$L(\phi) = \Delta\phi + pw(x)^{p-1}\phi - \phi = 0, \quad \phi \in L^{\infty}(\mathbb{R}^N) \qquad (1.8)$$

is a linear combination of the functions $\frac{\partial w}{\partial x_j}(x)$, $j = 1, \ldots, N$.

We want to solve the problem

$$\begin{cases} \varepsilon^2 \Delta \tilde{u} - V(x)\tilde{u} + \tilde{u}^p = 0 & \text{in} \quad \mathbb{R}^N \\ 0 < \tilde{u} \quad \text{in } \mathbb{R}^N & \tilde{u}(x) \to 0, \text{ as } |x| \to \infty. \end{cases} \qquad (1.9)$$

We fix a point $\xi \in \mathbb{R}^N$. Observe that $U_{\varepsilon,\xi}(y) := V(\xi)^{\frac{1}{p-1}} w\left(\sqrt{V(\xi)} \frac{y-\xi}{\varepsilon}\right)$, is a solution of the rescaled equation

$$\varepsilon^2 \Delta u - V(\xi)u + u^p = 0.$$

We will look for a solution of (1.9) such $u_\varepsilon(x) \approx U_{\varepsilon,\xi}(y)$ for some $\xi \in \mathbb{R}^N$. We define $w_\lambda = \lambda^{\frac{1}{p-1}} w(\sqrt{\lambda}x)$.

Let us observe that if $\tilde{u}$ satisfies (1.9), then $u(x) = \tilde{u}(\varepsilon z)$ satisfies the problem

$$\begin{cases} \Delta u - V(\varepsilon z)u + u^p = 0 & \text{in} \quad \mathbb{R}^N \\ 0 < u \quad \text{in } \mathbb{R}^N & u(x) \to 0, \text{ as } |x| \to \infty. \end{cases} \qquad (1.10)$$

Let $\xi' = \frac{\xi}{\varepsilon}$. We want a solution of (1.10) with the form $u(z) = w_\lambda(z - \xi') + \tilde{\phi}(z)$, with $\lambda = V(\xi)$ and $\tilde{\phi}$ being small compared with $w_\lambda(z - \xi')$.

1.3. Equation in terms of ϕ

Let $\phi(x) = \tilde{\phi}(x - \xi')$. Then ϕ satisfies the equation

$$\Delta_x[w_\lambda(x) + \phi(x)] - V(\xi + \varepsilon x)[w_\lambda(x) + \phi(x)] + [w_\lambda(x) + \phi(x)]^p = 0.$$

We can write this equation as

$$\Delta\phi - V(\xi)\phi + pw_\lambda^{p-1}(x)\phi - E + B(\phi) + N(\phi) = 0 \qquad (1.11)$$

where $E = (V(\xi + \varepsilon x) - V(\xi))w_\lambda(x)$, $B(\phi) = (V(\xi) - V(\xi + \varepsilon x))\phi$ and $N(\phi) = (w_\lambda + \phi)^p - w_\lambda^p - pw_\lambda^{p-1}\phi$.

We first consider the linear problem for $\lambda = V(\xi)$,

$$\begin{cases} L(\phi) = \Delta\phi - V(\xi + \varepsilon x)\phi + pw_\lambda(x)\phi = g - \sum_{i=1}^{N} c_i \frac{\partial w}{\partial x_i} \\ \int_{\mathbb{R}^N} \phi \frac{\partial w_\lambda}{\partial x_i} = 0, \quad i = 1, \ldots, N. \end{cases} \qquad (1.12)$$

The $c_i's$ are defined such that

$$\int_{\mathbb{R}^N} (L(\phi) - g) \frac{\partial w_\lambda}{\partial x_i} dx = 0, i = 1, \ldots, N \qquad (1.13)$$

which is equivalent to

$$\int_{\mathbb{R}^N} (L(\frac{\partial w_\lambda}{\partial x_i})\phi - g\frac{\partial w_\lambda}{\partial x_i})dx = 0, i = 1, \ldots, N \qquad (1.14)$$

Denoting

$$L_0(\phi) = \Delta\phi - V(\xi)\phi + pw_\lambda(x)\phi$$

and using the fact that

$$L_0(\frac{\partial w_\lambda}{\partial x_i}) = 0$$

we see that (1.14) can be further simplified as follows

$$\int_{\mathbb{R}^N} ((V(\xi) - V(\xi + \epsilon x))\frac{\partial w_\lambda}{\partial x_i}\phi - g\frac{\partial w_\lambda}{\partial x_i})dx = 0, i = 1, \ldots, N. \qquad (1.15)$$

Since

$$\int_{\mathbb{R}^N} \frac{\partial w_\lambda}{\partial x_i}\frac{\partial w_\lambda}{\partial x_j} = \int_{\mathbb{R}^N} (\frac{\partial w}{\partial x_1})^2 \delta_{ij}$$

we find that

$$c_i = \frac{\int_{\mathbb{R}^N}((V(\xi) - V(\xi + \epsilon x))\frac{\partial w_\lambda}{\partial x_i}\phi - g\frac{\partial w_\lambda}{\partial x_i})dx}{\int_{\mathbb{R}^N}(\frac{\partial w_\lambda}{\partial x_1})^2}, i = 1, \ldots, N. \qquad (1.16)$$

In the following, we shall solve the following:

Problem: Given $g \in L^\infty(\mathbb{R}^N)$, we want to find $\phi \in L^\infty(\mathbb{R}^N)$ solution to the problem (1.12)-(1.16).

1.4. *A priori estimates of a linear problem*

Let us assume that $V \in C^1(\mathbb{R}^N)$, $\|V\|_{C^1} < \infty$. We assume in addition that $|\xi| \leq M_0$ and $0 < \alpha \leq V$. Then we have

Proposition 1.2: *There exists ε_0, $C_0 > 0$ such that $\forall 0 < \varepsilon \leq \varepsilon_0$, $\forall |\xi| \leq M_0$, $\forall g \in L^\infty(\mathbb{R}^N) \cap C(\mathbb{R}^N)$, there exist a unique solution $\phi \in L^\infty(\mathbb{R}^N)$ to (1.12), $\phi = T[g]$ satisfies*

$$\|\phi\|_{C^1} \leq C_0\|g\|_\infty.$$

Proof: We divide the proof into two steps.

Step 1 - a priori estimates: We first obtain *a priori estimates* of the problem (1.12) on bounded domains $B_R(0)$: There exist R_0, ε_0, C_0 such that $\forall \varepsilon < \varepsilon_0$, $R > R_0$, $|\xi| \leq M_0$ such that $\forall \phi, g \in L^\infty$ solving $L(\phi) = g - \sum_i c_i \frac{\partial w_\lambda}{\partial x_i}$ in B_R, $\int_{B_R} \phi \frac{\partial w_\lambda}{\partial x_i} = 0$ and $\phi = 0$ on ∂B_R, we have

$$\|\phi\|_{C^1(B_R)} \leq C_0 \|g\|_\infty.$$

We prove first $\|\phi\|_\infty \leq C_0 \|g\|_\infty$. Assuming the opposite, then there exist sequences ϕ_n, g_n, $\varepsilon \to 0$, $R_n \to \infty$, $|\xi_n| \leq M_0$ such that

$$L(\phi_n) = g_n - \sum_i c_i^n \frac{\partial w_\lambda}{\partial x_i}.$$

The first fact is that $c_i^n \to 0$ as $n \to \infty$. This fact follows just after multiplying the equation against $\frac{\partial w_\lambda}{\partial x_i}$ and integrating by parts, as we did in (1.16).

We observe that if $\Delta \phi = g$ in B_2 then there exist C such that

$$\|\nabla \phi\|_{L^\infty(B_1)} \leq C[\|g\|_{L^\infty(B_2)} + \|\phi\|_{L^\infty(B_2)}]$$

where B_1 and B_2 are concentric balls. This implies that $\|\nabla \phi_n\|_{L^\infty(B)} \leq C$ a given bounded set B, $\forall n \geq n_0$. Hence passing to a subsequence we obtain $\phi_n \to \phi$ uniformly on compact sets, and $\phi \in L^\infty(\mathbb{R}^N)$. Observe that $\|\phi_n\|_\infty = 1$, and this implies that $\|\phi\|_\infty \leq 1$. We can also assume that up to a subsequence $\xi_n \to \xi_0$.

Since ϕ satisfies the equation $\Delta \phi - V(\xi_0)\phi + p w_{\lambda_0}^{p-1}(x)\phi = 0$, where $\lambda_0 = V(\xi_0)$, we have that $\phi \in \mathrm{Span}\left\{ \frac{\partial w_{\lambda_0}}{\partial x_1}, \ldots, \frac{\partial w_{\lambda_0}}{\partial x_N} \right\}$. Taking limits in the orthogonality condition (1.12) we obtain that $\int_{\mathbb{R}^N} \phi(w_{\lambda_0})_{\partial x_i} = 0$, $i = 1, \ldots, N$. This implies that $\phi = 0$ and hence $\|\phi_n\|_{L^\infty(B_M(0))} \to 0, \forall M < \infty$. Maximum principle yields that $\|\phi_n\|_{L^\infty(B_{R_n} \setminus B_{M_0})} \to 0$, since $|\phi_n| = o(1)$ on $\partial B_{R_n} \setminus B_{M_0}$ and $\|g_n\|_\infty \to 0$. Therefore we arrive at $\|\phi_n\|_\infty \to 0$, which is a contradiction. This implies that $\|\phi\|_{L^\infty(B_R)} \leq C_0 \|g\|_{L^\infty(B_R)}$ uniformly on large R. The C^1 estimate follows from elliptic local boundary estimates for elliptic operators.

Step 2 - Existence: Recall that $g \in L^\infty$. We want to solve (1.12). We claim that solving (1.12) is equivalent to finding

$$\phi \in X = \{\psi \in H_0^1(B_R) : \int \psi \frac{\partial w_\lambda}{\partial x_i} = 0, \, i = 1, \ldots, N\}$$

such that

$$\int \nabla \phi \nabla \psi + \int V(\xi + \varepsilon x)\phi \psi - p w^{p-1} \phi \psi + \int g \psi = 0, \quad \forall \psi \in X.$$

Take general $\Psi \in H_0^1$. We can decompose into $\Psi = \psi - \sum_i \alpha_i \frac{\partial w_\lambda}{\partial x_i}$, with $\alpha_i = \frac{\int \Psi \frac{\partial w_\lambda}{\partial x_i}}{\int (\frac{\partial w_\lambda}{\partial x_i})^2}$. We have

$$-\int \Delta(\sum_i \alpha_i \frac{\partial w_\lambda}{\partial x_i})\nabla\phi + \int V(\xi)(\sum_i \alpha_i(\frac{\partial w_\lambda}{\partial x_i})\phi - pw^{p-1}(\sum_i \alpha_i \frac{\partial w_\lambda}{\partial x_i})\phi = 0$$

which implies that

$$\int \nabla\phi\nabla\Psi + \int V(\xi)\phi\Psi - pw^{p-1}\phi\Psi$$

$$-\int (V(\xi) - V(\xi + \varepsilon x))(\Psi - \sum_i \alpha_i \frac{\partial w_\lambda}{\partial x_i}) + \int g(\Psi - \sum_i \alpha_i \frac{\partial w_\lambda}{\partial x_i})$$

$$= \int [(V(\xi + \varepsilon x) - V(\xi))\phi + g](\Psi - \sum_i \alpha_i \frac{\partial w_\lambda}{\partial x_i}).$$

Let $\Pi_X(\Psi) = \sum_i \alpha_i \frac{\partial w_\lambda}{\partial x_i}$. Then the above integral equals

$$\int \Pi_X([(V(\xi + \varepsilon x) - V(\xi))\phi + g]\phi)\Psi.$$

This implies that

$$-\Delta\phi + V(\xi)\phi - pw^{p-1}\phi + \Pi_X([(V(\xi + \varepsilon x) - V(\xi))\phi + g]\phi) = 0.$$

The problem is formulated weakly as

$$\int \nabla\phi\nabla\psi + \int (V(\xi + \varepsilon x) - pw^{p-1})\phi\psi + \int g\psi = 0, \phi \in X, \forall \psi \in X$$

which can be written as $\phi = A[\phi] + \tilde{g}$, where A is a compact operator. The *a priori* estimate implies that the only solution when $g = 0$ of this equation is $\phi = 0$. We conclude existence by Fredholm alternative. Finally we let $R \to +\infty$ and obtain the existence in the whole space, thanks to the a priori estimate in Step 1. $\qquad\square$

Next we consider the assembly of multiple spikes. We look for a solution of (1.10) which near $x_j = \xi'_j = \xi_j/\varepsilon, j = 1, \ldots, k$ looks like $v(x) \approx w_{\lambda_j}(x - \xi'_j)$, $\lambda_j = V(\xi_j)$, where $w_\lambda = \lambda^{1/(p-1)}w(\sqrt{\lambda}y)$.

Let $\xi_1, \xi_2, \ldots, \xi_k \in \mathbb{R}^N$ be such that $|\xi'_j - \xi'_l| \gg 1$, if $j \neq l$. We look for a solution $v(x) \approx \sum_{j=1}^k w_{\lambda_j}(x - \xi'_j)$, $\lambda_j = V(\xi_j)$. We assume $V \in C^2(\mathbb{R}^N)$ and $\|V\|_{C^2} < \infty$, $0 < \alpha \le V$. We use the notation $W_j = w_{\lambda_j}(x - \xi'_j)$, $\lambda_j = V(\xi_j)$ and $W = \sum_{j=1}^k W_j$.

Setting $v = W + \phi$, then ϕ solves the problem

$$\Delta\phi - V(\varepsilon x)\phi + pW^{p-1}\phi + E + N(\phi) = 0 \tag{1.17}$$

where

$$E = \Delta W - VW + W^p, \quad N(\phi) = (W + \phi)^p - W^p - pW^{p-1}\phi.$$

Observe that $\Delta W = \sum_j \Delta W_j = \sum_j \lambda_j W_j - W_j^p$. So we can write

$$E = \sum_j (\lambda_j - V(\varepsilon x))W_j + (\sum_j W_j)^p - \sum_j W_j^p.$$

Our next objective is to solve the approximate linearized projected problem.

1.5. *Linearized (projected) problem*

We use the following notation $Z_j^i = \frac{\partial W_j}{\partial x_i}$. The linearized projected problem is the following

$$\Delta\phi - V(\varepsilon x)\phi + pW^{p-1}\phi + g = \sum_{i,j} c_j^i Z_j^i, \tag{1.18}$$

with the orthogonality condition $\int \phi Z_j^i = 0$, $\forall i, j$. The Z_j^i's are "nearly orthogonal" if the centers ξ_j' are far away one to each other. The c_j^i's are, by definition, the solution of the linear system

$$\int_{\mathbb{R}^N} (\Delta\phi - V(\varepsilon x)\phi + pW^{p-1}\phi + g)Z_{j_0}^{i_0} = \sum_{i,j} c_j^i \int_{\mathbb{R}^N} Z_j^i Z_{j_0}^{i_0},$$

for $i_0 = 1, \ldots, N$, $j_0 = 1, \ldots, k$. The c_j^i's are indeed uniquely determined provided that $|\xi_l' - \xi_j'| > R_0 \gg 1$, because the matrix with coefficients $\alpha_{i,j,i_0,j_0} = \int Z_j^i Z_{j_0}^{i_0}$ is "nearly diagonal", which means

$$\alpha_{i,j,i_0,j_0} = \begin{cases} \frac{1}{N}\int |\nabla W_j|^2 & \text{if } (i,j) = (i_0, j_0), \\ o(1) & \text{if not.} \end{cases}$$

Moreover by a similar argument leading to (1.15) we have

$$|c_{j_0}^{i_0}| \leq C\sum_{i,j}\int |\phi|[|\lambda_j - V| + p|W^{p-1} - W_j^{p-1}|]|Z_j^i| + \int |g||Z_j^i|$$

$$\leq C(\|\phi\|_\infty + \|g\|_\infty)$$

with C is uniform for large R_0. Furthermore if we rescale $x = \xi' + y$, we get

$$|(\lambda_j - V(\varepsilon x))Z_j^i| \leq |(V(\xi_j) - V(\xi_j + \varepsilon y))||\frac{\partial w_{\lambda_j}}{\partial y_i}| \leq C\varepsilon e^{-\frac{\sqrt{\alpha}}{2}|y|},$$

because $|\frac{\partial w_{\lambda_j}}{\partial y_i}| \leq Ce^{-|y|\sqrt{\lambda_j}}|y|^{-(N-1)/2}$. Observe also that

$$|(W^{p-1} - W_j^{p-1})Z_j^i| = |((1 - \sum_{l \neq j} \frac{W_l}{W_j})^{p-1} - 1)|W_j^{p-1}Z_j^i.$$

We estimate the interactions at each spike in two regions. Observe that if $|x - \xi_j'| < \delta_0 \min_{j_1 \neq j_2} |\xi_{j_1}' - \xi_{j_2}'|$, then

$$\frac{W_l(x)}{W_j(x)} \approx \frac{e^{-\sqrt{\lambda_l}|x-\xi_l'|}}{e^{-\sqrt{\lambda_j}|x-\xi_j'|}} < \frac{e^{-\sqrt{\lambda_l}|x-\xi_l'|}}{e^{-\sqrt{\lambda_j}\delta_0 \min_{j_1 \neq j_2}|\xi_{j_1}' - \xi_{j_2}'|}}.$$

If $\delta_0 \ll 1$ but fixed, we conclude that $e^{-\sqrt{\lambda_l}|\xi_j' - \xi_l'| + \delta_0(\sqrt{\lambda_l} - \sqrt{\lambda_j})\min_{j_1 \neq j_2}|\xi_{j_1}' - \xi_{j_2}'|} < e^{-\rho \min_{j_1 \neq j_2}|\xi_{j_1}' - \xi_{j_2}'|} \ll 1$. Thus we conclude that if $|x - \xi_j'| < \delta_0 \min_{j_1 \neq j_2} |\xi_{j_1}' - xi_{j_2}'|$ then

$$|(W^{p-1} - W_j^{p-1})Z_j^i| \leq e^{-\rho \min_{j_1 \neq j_2}|\xi_{j_1}' - \xi_{j_2}'|}e^{-\frac{\alpha}{2}|x-\xi_j'|}.$$

On the other hand if $|x - \xi_j'| > \delta_0 \min_{j_1 \neq j_2} |\xi_{j_1}' - \xi_{j_2}'|$, then

$$|(W^{p-1} - W_j^{p-1})Z_j^i \leq C|Z_j^i| \leq Ce^{-\rho \min_{j_1 \neq j_2}|\xi_{j_1}' - \xi_{j_2}'|}e^{-\frac{\alpha}{2}|x-\xi_j'|}.$$

As a conclusion we obtain the following estimate

$$|c_{j_0}^{i_0}| \leq C(\varepsilon + e^{-\rho \min_{j_1 \neq j_2}|\xi_{j_1}' - \xi_{j_2}'|})\|\phi\|_\infty + \|g\|_\infty. \tag{1.19}$$

Lemma 1.1: *Given $k \geq 1$, there exist R_0, C_0, ε_0 such that for all points ξ_j' with $|\xi_{j_1}' - \xi_{j_2}'| > R_0$, $j = 1, \ldots, k$ and all $\varepsilon < \varepsilon_0$ then exist a unique solution ϕ to the linearized projected problem with*

$$\|\phi\|_\infty \leq C_0\|g\|_\infty.$$

Proof: As before we first prove the *a priori* estimate $\|\phi\|_\infty \leq C_0\|g\|_\infty$. If not there exist $\varepsilon_n \to 0$, $\|\phi_n\|_\infty = 1$, $\|g_n\| \to 0$, $\xi_j'^n$ with $\min_{j_1 \neq j_2} |\xi_{j_1}'^n - \xi_{j_2}'^n| \to \infty$. We denote $W_n = \sum_j W_{j_n}$, and we have

$$\Delta\phi_n - V(\varepsilon_n x)\phi_n + pW_n^{p-1}\phi_n + g_n = \sum_{i,j}(c_j^i)_n(z_j^i)_n.$$

Our first observation is that $(c_j^i)_n \to 0$ (which follows from the same estimate for $c_{j_0}^{i_0}$). Next we claim that $\forall R > 0$ $\|\phi_n\|_{L^\infty(B(\xi_j'^n, R))} \to 0$, $j = 1, \ldots, k$. If not, there exist j_0 $\|\phi_n\|_{L^\infty(B(\xi_{j_0}'^n, R))} \geq \gamma > 0$. We denote $\tilde{\phi}_n(y) := \phi_n(\xi_{j_0}'^n + y)$. We have $\|\tilde{\phi}_n\|_{L^\infty(B(0,R))} \geq \gamma > 0$. Since $|\Delta\tilde{\phi}_n| \leq C$, $\|\tilde{\phi}_n\|_\infty \leq 1$. This implies that $\|\nabla\tilde{\phi}_n\| \leq C$. Passing to a subsequence we may assume

$\tilde{\phi}_n \to \tilde{\phi}$ uniformly on compacts sets. Observe that also $V(\varepsilon_n(\xi_{j_0}'^n + y)) = V(\varepsilon_n \xi_{j_0}'^n) + O(\varepsilon_n |y|) \to \lambda_{j_0}$ over compact sets and $W_n(\xi_{j_0}'^n + y) \to W_{\lambda_{j_0}}(y)$ uniformly on compact sets. This implies that $\tilde{\phi}$ is a solution of the problem

$$\Delta\tilde{\phi} - \lambda_{j_0}\tilde{\phi} + pw_{\lambda_0}^{p-1}p\tilde{} - 1 = 0, \quad \int \tilde{\phi}\frac{\partial W_{\lambda_{j_0}}}{\partial y_i}dy = 0, i = 1,\ldots,N.$$

Nondegeneracy of $w_{\lambda_{j_0}}$ implies that $\tilde{\phi} = \sum_i \alpha_i \frac{\partial w_{\lambda_{j_0}}}{\partial y_i}$. The orthogonality condition implies that $\alpha_i = 0, \forall i = 1,\ldots,N$. This implies that $\tilde{\phi} = 0$ but $\|\tilde{\phi}\|_{L^\infty(B(0,R))} \geq \gamma > 0$, a contradiction.

Now we prove: $\|\phi_n\|_{L^\infty}(\mathbb{R}^N \setminus \cup_n B(\xi_j'^n, R)) \to 0$, provided that $R \gg 1$ and fixed so that $\phi_n \to 0$ in the sense of $\|\phi_n\|_\infty$ (again a contradiction). We will denote $\Omega_n = \mathbb{R}^N \setminus \cup_n B(\xi_j'^n, R)$. For $R \gg 1$ the equation for ϕ_n has the form

$$\Delta\phi_n - Q_n\phi_n + g_n = 0$$

where $Q_n = V(\varepsilon x) - pW_n^{p-1} \geq \frac{\alpha}{2} > 0$ for some R sufficiently large (but fixed).

Let us take for $\sigma^2 < \alpha/2$

$$\bar{\phi} = \delta \sum_j e^{\sigma|x-\xi_j'^n|} + \mu_n.$$

We denote $\varphi(y) = e^{\sigma|y|}$, $r = |y|$. Observe that $\Delta\varphi - \alpha/2\varphi = e^{\sigma|y|}(\sigma^2 + \frac{N-1}{|y|} - \alpha/2) < 0$ if $|y| > R \gg 1$. Then

$$-\Delta\bar{\phi} + Q_n\bar{\phi} - g_n > -\Delta\bar{\phi} + \frac{\alpha}{2}\bar{\phi} - \|g_n\|_\infty > \frac{\alpha}{2}\mu_n - \|g_n\|_\infty > 0 \quad (1.20)$$

if we choose $\mu_n \geq \|g_n\|_\infty\frac{2}{\alpha}$. In addition we take $\mu_n = \sum_j \|\phi_n\|_{L^\infty(B(\xi_j^n,R))} + \|g_n\|_\infty\frac{2}{\alpha}$. Maximum principle implies that $\phi_n(x) \leq \bar{\phi}$ for all $x \in \Omega_n$. Taking $\delta \to 0$ this implies that $\phi_n(x) \leq \mu_n$, for all $x \in \Omega_n$. It is also true that $|\phi_n(x)| \leq \mu_n$ for all $x \in \Omega_n^c$, and this implies that $\|\phi_n\|_{L^\infty(\mathbb{R}^N)} \to 0$. $\square$

Remark: If in addition we have the following decay for the error

$$\theta_n = \|g_n\left(\sum_j e^{-\rho|x-\xi_j'^n|}\right)^{-1}\|_\infty \to 0$$

with $\rho < \alpha/2$, then we can use as a barrier function

$$\bar{\phi} = \delta \sum_j e^{\sigma|x-\xi_j'^n|} + \mu_n \sum_j e^{-\rho|x-\xi_j'^n|}$$

with $\mu_n = e^{\rho R} \sum_j \|\phi_n\|_{L^\infty(B(\xi_j'^n, R))} + \theta_n$. It is easy to see that $\bar{\phi}$ is a super solution of the equation in $(\cup_j B(\xi_j, R))^c$ and we have $|\phi_n| \leq \bar{\phi}$. Letting $\delta \to 0$, we get $|\phi_n(x)| \leq \mu_n \sum_j e^{-\rho|x - \xi_j'^n|}$. As a conclusion we also get the *a priori* estimate

$$\left\|\phi \left(\sum_{j=1}^{k} e^{-\rho|x - \xi_j'|}\right)^{-1}\right\|_\infty \leq C\|g\left(\sum_{j=1}^{k} e^{-\rho|x - \xi_j'|}\right)^{-1}\|_\infty$$

provided that $0 \leq \rho < \alpha/2$, $|\xi_{j_1}' - \xi_{j_2}'| > R_0 \gg 1$, $\varepsilon < \varepsilon_0$.

We now give the proof of existence.

Proof: Let g be compactly supported smooth functions. The weak formulation for

$$\Delta\phi - V(\varepsilon x)\phi + pW^{p-1}\phi + g = \sum_{i,j} c_j^i Z_j^i, \quad \int \phi Z_j^i = 0, \forall i, j \quad (1.21)$$

is to find $\phi \in X = \{\phi \in H^1(\mathbb{R}^N) : \int \phi Z_j^i = 0, \forall i, j\}$ such that

$$\int_{\mathbb{R}^N} \nabla\phi\nabla\psi + V\phi\psi - pW^{p-1}\phi\psi - g\psi = 0, \quad \forall\psi \in X. \quad (1.22)$$

Assume ϕ solves (1.21). For $g \in L^2$, we decompose $g = \tilde{g} + \Pi[g]$ where $\int \tilde{g}Z_j^i = 0$ for all i, j, and Π is the orthogonal projection of g onto the space spanned by the Z_j^i's.

Let $\psi \in H^1(\mathbb{R}^N)$. We now use $\psi - \Pi[\psi]$ as a test function in (1.22). Then if $\varphi \in C_c^\infty(\mathbb{R}^N)$, then we have

$$\int_{\mathbb{R}^N} \nabla\varphi\nabla(\Pi[\psi]) = -\int_{\mathbb{R}^N} \Delta\varphi\Pi[\psi] = -\int_{\mathbb{R}^N} \Pi[\Delta\varphi]\psi. \quad (1.23)$$

On the other hand, we have $\Pi[\Delta\varphi] = \sum_{i,j} \alpha_{i,j} Z_j^i$, where

$$\sum \alpha_{i,j} \int Z_{i,j} Z_{i_0, j_0} = \int \Delta\varphi Z_{i_0}^{j_0} = \int \varphi\Delta Z_{i_0}^{j_0}. \quad (1.24)$$

Then $\|\Pi[\Delta\varphi]\|_{L^2} \leq C\|\varphi\|_{H^1}$. By density argument, it is also true for $\varphi \in H^1$ where $\Delta\varphi \in H^{-1}$. Therefore

$$\int \nabla\phi\nabla\psi + \int (V\phi - pW^{p-1}\phi - g)\psi = \int \Pi(V\phi - pW^{p-1}\phi + g)\psi. \quad (1.25)$$

It follows that ϕ solves in weak sense

$$-\Delta\phi + V\phi - pW^{p-1}\phi - g = \Pi[-\Delta\phi + V\phi - pW^{p-1}\phi - g] \quad (1.26)$$

and $\Pi[-\Delta\phi + V\phi - pW^{p-1}\phi - g] = \sum_{i,j} c_i^j Z_i j$. Therefore by definition ϕ solves (1.22) implies that ϕ solves (1.26). Classical regularity gives that this weak solution is solution of (1.26) in strong sense, in particular $\phi \in L^\infty$ so that

$$\|\phi\|_\infty \le C\|g\|_\infty. \tag{1.27}$$

Now we give the proof of existence for (1.21). We take g compactly supported. The equation (1.26) can be written in the following way (using Riesz theorem):

$$\langle\phi, \psi\rangle_{H^1} + \langle B[\phi], \psi\rangle_{H^1} = \langle\tilde{g}, \psi\rangle_{H^1} \tag{1.28}$$

or $\phi + B[\phi] = \tilde{g}$, $\phi \in X$. We claim that B is a compact operator. Indeed if $\phi_n \rightharpoonup 0$ in X, then $\phi_n \to 0$ in L^2 over compacts and

$$|\langle B[\phi_n], \psi\rangle| \le |\int pW^{p-1}\phi_n\psi| \le (\int pw^{p-1}\phi_n^2)^{1/2}(\int pW^{p-1}\psi^2)^{1/2} \tag{1.29}$$

which yields

$$|\langle B[\phi_n], \psi\rangle| \le c(\int pW^{p-1}\phi_n^2)^{1/2}\|\psi\|_{H^1}. \tag{1.30}$$

Take $\psi = B[\phi_n]$, which implies

$$\|B[\phi_n]\|_{H^1} \le c(\int pW^{p-1}\phi_n^2)^{1/2} \to 0. \tag{1.31}$$

This gives that B is a compact operator.

Now we prove existence with the aid of Fredholm alternative. Problem (1.21) is solvable if for $\tilde{g} = 0$ the only solution to (1.22) is $\phi = 0$. But $\phi + B[\phi] = 0$ implies solve (1.21) (strongly) with $g = 0$. This implies $\phi \in L^\infty$, and the *a priori* estimate implies $\phi = 0$. Considering $g\Xi_{B_R(0)}$ we conclude that

$$\|\phi_R\|_\infty \le \|g\|_\infty. \tag{1.32}$$

Taking $R \to \infty$ then along a subsequence $\phi_R \to \phi$ uniform over compacts we obtain a solution to (1.21). $\qquad\square$

Next, we want to study the dependence and regularity of the solution with respect to the parameters. Let $g \in L^\infty$. We denote $\phi = T_{\xi'}[g]$, where $\xi' = (\xi_1', \ldots, \xi_k')$. We want to analyze derivatives $\partial_{\xi_{ji}'} T_{\xi'}[g]$. We know that $\|T_{\xi'}[g]\| \le C_0\|g\|_\infty$. First, we make a formal differentiation. We denote $\Phi = \frac{\partial\phi}{\partial\xi_{i_0 j_0}'}$.

We have $\Delta\phi - V\phi + pW^{p-1}\phi + g = \sum_{i,j} c_j^i Z_j^i$ and $\int \phi Z_j^i = 0$, for all i, j. Formal differentiation yields

$$\Delta\Phi - V\Phi + pW^{p-1}\Phi + +\partial_{\xi_{i_0 j_0}}(W^{p-1})\phi - \sum_{i,j} c_j^i \partial_{\xi_{i_0 j_0}} Z_i^j = \sum_{i,j} \tilde{c}_j^i Z_j^i \quad (1.33)$$

where formally $\tilde{c}_i^j = \partial_{\xi_{i_0 j_0}} c_i^j$. The orthogonality conditions is reduced to

$$\int_{\mathbb{R}^N} \Phi Z_j^i = \begin{cases} 0 & \text{if } j \neq j_0 \\ -\int \phi \partial_{\xi_{i_0 j_0}} Z_{j_0}^i & \text{if } j = j_0. \end{cases} \quad (1.34)$$

Let us define $\tilde{\Phi} = \Phi - \sum_{i,j} \alpha_{i,j} Z_j^i$. We want $\int \tilde{\Phi} Z_j^i = 0$, for all i, j. We need

$$\sum_{i,j} \alpha_{i,j} \int Z_j^i Z_{\bar{j}}^{\bar{i}} = \begin{cases} 0 & \text{if } \bar{j} \neq j_0 \\ -\int \phi \partial_{\xi_{i_0 j_0}} Z_{j_0}^i & \text{if } \bar{j} = j_0. \end{cases} \quad (1.35)$$

The system has a unique solution and $|\alpha_{i,j}| \leq C\|\phi\|_\infty$ (since the system is almost diagonal). So we have the condition $\int \tilde{\Phi} Z_j^i = 0$, for all i, j. We add to the equation the term $\sum_{i,j} \alpha_{i,j}(\Delta - V + pW^{p-1})Z_j^i$, so $\tilde{\Phi}$ satisfies the equation $\Delta\phi - V\phi + pW^{p-1}\phi + g = \sum_{i,j} c_j^i Z_j^i$

$$\Delta\tilde{\Phi} - V\tilde{\Phi} + pW^{p-1}\tilde{\Phi} + \partial_{\xi_{i_0 j_0}}(W^{p-1})\phi - \sum_{i,j} c_j^i \partial_{\xi_{i_0 j_0}} Z_i^j$$

$$= \sum_{i,j} \tilde{c}_j^i Z_j^i - \sum_{i,j} \alpha_{i,j}(\Delta - V + pW^{p-1})Z_j^i. \quad (1.36)$$

This implies $\|\tilde{\Phi}\| \leq C(\|h\| + \|g\|) \leq C\|g\|_\infty$ and hence $\|\Phi\| \leq C\|g\|_\infty$.

The above formal procedure can be made rigorous by performing the analysis discretely, namely we consider solutions corresponding to ξ and $\xi + h$ respectively. Then we consider the quotient and pass the limit in h. We omit the details. In conclusion the map $\xi \to \partial_\xi \phi$ is well defined and continuous (into L^∞). Besides we also have $\|\partial_\xi \phi\|_\infty \leq C\|g\|_\infty$, and this implies

$$\|\partial_\xi T_\xi[\phi]\| \leq C\|g\|. \quad (1.37)$$

1.6. *Nonlinear projected problem*

Consider now the nonlinear projected problem

$$\Delta\phi - V\phi + pw^{p-1}\phi + E + N(\phi) = \sum_{i,j} c_j^i Z_j^i, \quad \int \phi Z_i^j = 0, \forall i, j. \quad (1.38)$$

We solve this by fixed point. We have $\phi = T(E + N(\phi)) =: M(\phi)$. We define $\Lambda = \{\phi \in C(\mathbb{R}^N) \cap L^\infty(R^N) : \|\phi\|_\infty \leq M\|E\|_\infty\}$. Remember that

$E = \sum_i (\lambda_j - V(\varepsilon x)) W_j + (\sum_j W_j)^p - \sum_j W_j^p$. Observe that

$$|E| \le \varepsilon \sum_i e^{-\sigma|x-\xi_j'|} + ce^{-\delta_0 \min_{j_1 \ne j_2} |\xi_{j_1}' - \xi_{j_2}'|} \sum_j e^{-\sigma|x-\xi_j'|} \qquad (1.39)$$

so, for existence we have $\|E\| \le C[\varepsilon + e^{-\delta_0 \min_{j_1 \ne j_2} |\xi_{j_1}' - \xi_{j_2}'|}] =: \rho$ (see that ρ is small). Contraction mapping implies there exists a unique solution $\phi = \Phi(\xi)$ and $\|\Phi(\xi)\| \le M\rho$. The proof is standard and hence omitted.

1.7. Differentiability in ξ' of $\Phi(\xi')$

As before the solutions obtained for the nonlinear projected problem has more regularity. In fact, we can write the equation for Φ as

$$\Phi - T_\xi'(E_\xi' + N_\xi'(\phi)) = A(\Phi, \xi') = 0. \qquad (1.40)$$

If $(D_\Phi A)(\Phi(\xi'), \xi')$ is invertible in L^∞, then $\Phi(\xi')$ turns out to be of class C^1. This is a consequence of the fixed point characterization, i.e., $D_\Phi A(\Phi(\xi'), \xi') = I + o(1)$ (the order $o(1)$ is a direct consequence of fixed point characterization). Then it is invertible. Contraction mapping theorem yields the existence of C^1 derivative of $A(\Phi, \xi')$ in (ϕ, ξ'). This implies $\Phi(\xi')$ is C^1. With a little bit of more work we can show that $\|D_\xi'\Phi(\xi')\| \le C\rho$ (just using the derivative given by the implicit function theorem).

1.8. Solving the reduced problem: Direct method

By (1.38), to solve (1.17), we need to find ξ' such that the reduced problem

$$c_j^i = 0, \forall i, j \qquad (1.41)$$

to get a solution to the original problem (1.10). There are two ways to solve the reduced problem (1.41): the first one is the direct method, and the second one is the variational reduction method. We describe the first method first by proving the following

Theorem 1: *(Oh [54]) Assume that $\xi_j^0, j = 1, ..., k$ are k distinct non-degenerate critical points of V. Then there exist a solution u_ε to the original problem with*

$$u_\varepsilon(x) \approx \sum_{j=1}^k w_{V(\xi_j^\varepsilon)}(x - \xi_j^\varepsilon/\varepsilon), \quad \xi_j^\varepsilon \to \xi_j^0.$$

Proof: To solve the problem (1.41) we first obtain the asymptotic formula for c_j^i. To this end we multiply the equation (1.38) by $Z_{j_0}^{i_0}$ and integrate by parts. We obtain

$$\int_{\mathbb{R}^N} Z_j^i Z_{j_0}^{i_0} c_j^i = \int_{\mathbb{R}^N} (V(\xi_j + \epsilon x) - V(\xi_j)) w_{\xi_j} Z_{j_0}^{i_0} + O(\epsilon^2)$$

and thus

$$c_{j_0}^{i_0} \sim \partial_{i_0} V(\xi_j^0) + O(\epsilon).$$

The nondegeneracy of the critical point $\nabla V(\xi_j^0)$ and implicit function theorem yields the existence of $\xi_j = \xi_j^0 + O(\epsilon)$ such that (1.41) holds. $\square$

The direct method can be used to construct multiple spike solutions for problems *without variational structure*, such as Gierer-Meinhardt system. For this application, we refer to [62].

1.9. *Solving the reduced problem: Variational reduction*

If the problem concerned has a variational structure, it is more appropriate to use a variational reduction method to solve (1.41). This method gives much stronger results under very weak assumptions.

We now describe the procedure that we call Variational Reduction in which the problem of finding ξ' with $c_j^i = 0$, for all i, j, is equivalent to finding a critical point of a reduced functional of ξ'.

Define an energy functional

$$J(v) = \frac{1}{2} \int_{\mathbb{R}^N} |\nabla v|^2 + V(\varepsilon x) v^2 - \frac{1}{p+1} \int_{\mathbb{R}^{N+1}} v_+^{p+1} \tag{1.42}$$

where $v \in H^1(\mathbb{R}^N)$ and $1 < p < \frac{N+2}{N-2}$. Since p is subcritical, by standard elliptic regularity arguments and Maximum Principle v is a solution of the problem

$$\Delta v - V v + v^p = 0, v \to 0 \tag{1.43}$$

if and only if $v \in H^1(\mathbb{R}^N)$ and $J'(v) = 0$. Observe that $\langle J'(v), \varphi \rangle = \int \nabla v \nabla \varphi + V v \varphi - v_+^p \varphi$.

We will prove the following Variational Reduction Principle:

Theorem 2: $v = W_{\xi'} + \phi(\xi')$ *is a solution of the original problem (for $\rho \ll 1$) if and only if*

$$\partial_{\xi'} J(W_{\xi'} + \phi(\xi'))|_{\xi' = \xi'_*} = 0. \tag{1.44}$$

Proof: Indeed, observe that $v(\xi') := W_{\xi'} + \phi(\xi')$ solves the problem $\Delta v(\xi') - V(\varepsilon x)v(\xi') + v(\xi')^p = \sum_{i,j} c_j^i Z_j^i$ and also that

$$\partial_{\xi'_{j_0 i_0}} J(v(\xi')) = \langle J'(v(\xi')), \partial_{\xi'_{j_0 i_0}} v(\xi') \rangle = -\sum_{j,i} c_j^i \int Z_j^i \partial_{\xi'_{j_0 i_0}} v$$

$$= -\sum_{i,j} c_j^i \int Z_i^j (\partial_{\xi'_{j_0 i_0}} W_{\xi'} + \partial_{\xi'_{j_0 i_0}} \phi(\xi')). \qquad (1.45)$$

Recall that $W_{\xi'} = \sum_{j=1}^k w_{\lambda_j}(x - \xi'_j)$,

$$\partial_{\xi'_{j_0 i_0}} W_\xi = \partial_{\xi'_{j_0 i_0}} w_{\lambda_{j_0}(\xi')}(x - \xi'_j) = (\partial_\lambda w_\lambda(x - \xi'_{j_0}))|_{\lambda = \lambda_{j_0}} - \partial_{x_{i_0}} w_{\lambda_{j_0}}(x - \xi'_{j_0})$$

$$= O(e^{-\delta|x - \xi'_0|})o(\varepsilon) - Z_{j_0 i_0}(x). \qquad (1.46)$$

This is because $\partial_\lambda w_\lambda = O(e^{-\delta|x - \xi'_0|})$. On the other hand since $\int Z_i^j \phi(\xi') = 0$ we have

$$\int Z_i^j \partial_{\xi'_{j_0 i_0}} \phi(\xi') = -\int \phi(\xi') \partial_{\xi'_{j_0 i_0}} Z_i^j$$

which is small thanks to the fact that $|\phi| \leq C\rho e^{-\delta|x - \xi'_{j_0}|}$. Finally, observe that

$$-\int Z_j^i (\partial_{\xi'_{j_0 i_0}} W'_\xi + \partial_{\xi'_{j_0 i_0}} \phi) = \int Z_j^i Z_{j_0}^{i_0} + O(\rho). \qquad (1.47)$$

The matrix of these numbers is invertible provided $\rho \ll 1$. □

We now discuss several applications of the reduction principle.

Theorem 3: *(del Pino and Felmer [15]) Assume that there exists an open, bounded set $\Lambda \subset \mathbb{R}^N$ such that*

$$\inf_{\partial \Lambda} V > \inf_\Lambda V, \qquad (1.48)$$

then there exist a solution to the original problem, v_ε with $v_\varepsilon(x) = w_{V(\xi_\varepsilon)}((x - \xi_\varepsilon)/\varepsilon) + o(1)$ and $V(\xi_\varepsilon) \to \min_\Lambda V$, $\xi = \xi_\varepsilon$.

Theorem 4: *(del Pino-Felmer [16]) Assume that $\Lambda_1, \ldots, \Lambda_k$ are disjoint bounded sets with*

$$\inf_{\Lambda_j} V < \inf_{\partial \Lambda_j} V, j = 1, \ldots, k.$$

Then there exist a solution u_ε to the original problem with

$$u_\varepsilon(x) \approx \sum_{j=1}^k w_{V(\xi_j^\varepsilon)}(x - \xi_j^\varepsilon/\varepsilon), \quad \xi_j^\varepsilon \in \Lambda_j$$

and $V(\xi_j^\varepsilon) \to \inf_{\Lambda_j} V$. The same result holds if the minimum is replaced by maximum.

Theorem 5: *(Kang-Wei [39]) Let Γ be a bounded open set such that*

$$\max_{\Gamma} V(x) > \max_{\partial\Gamma} V(x).$$

Then for any positive integer K there exists a solution u_ϵ such that

$$u_\varepsilon(x) \approx \sum_{j=1}^{k} w_{V(\xi_j^\varepsilon)}(x - \xi_j^\varepsilon/\varepsilon), \quad \xi_j^\varepsilon \in \Lambda, V(\xi_j^\varepsilon) \to \max_{\Lambda} V(x).$$

Proof: Assume that $j = 1$ first so that $v(\xi') = W_{\xi'} + \phi(\xi')$. Then we can compute the reduced energy as follows:

$$J(v(\xi')) = J(W_{\xi'} + \phi(\xi')) + \langle J'(W_\xi' + \phi), -\phi\rangle + \frac{1}{2}J''(W_\xi' + (1-t)\phi)[\phi]^2. \quad (1.49)$$

(This follows from Taylor expansion of the function $\alpha(t) = J(W_{\xi'} + (1 - t)\phi)$.) Observe that $\langle J'(W_\xi' + \phi), -\phi\rangle = \sum_{i,j} c_j^i \int Z_i^j \phi = 0$. Also observe that

$$J''(W_\xi' + (1-t)\phi)[\phi]^2 = \int |\nabla\phi|^2 + V(\varepsilon x)\phi^2 - p(W_\xi' + (1-t)\phi)\phi^2 = O(\varepsilon^2) \quad (1.50)$$

uniformly on ξ' because $\nabla\phi, \phi = O(\varepsilon e^{-\delta|x-\xi'|})$. We call $\Phi(\xi) := J(v(\xi')) = J(W_{\xi'}) + O(\varepsilon^2)$, and

$$J(W_{\xi'}) = \frac{1}{2}\int |\nabla W_{\xi'}|^2 + V(\xi)W_{\xi'}^2 - \frac{1}{p+1}\int W_{\xi'}^{p+1} + \int (V(\varepsilon x) - V(\xi'))W_{\xi'}^2. \quad (1.51)$$

Taking $\lambda = V(\xi)$, we have that

$$\int |\nabla w_\lambda(x)|^2 = \lambda^{-N/2}\int |\nabla w(\lambda^1/2x)|^2 \lambda^{1+2/(p-1)} \lambda^{N/2} dx$$

$$= \lambda^{-N/2+p+1/p-1}|\nabla w(y)|^2 dy \quad (1.52)$$

and

$$\lambda \int w_\lambda^2(x) = \lambda^{-N/2p+1/p-1}\int w(y)^{p+1} dy. \quad (1.53)$$

This implies that

$$\frac{1}{2}\int |\nabla W_{\xi'}|^2 + V(\xi')W_{\xi'}^2 - \frac{1}{p+1}\int W_{\xi'}^{p+1} = V(\xi')^{p+1/p-1-N/2} c_{p,N} \quad (1.54)$$

and we also have

$$\int (V(\varepsilon x) - V(\xi'))w_\lambda(x - \xi')^2 = O(\varepsilon) \quad (1.55)$$

uniformly in ξ'.

In summary, we have the following asymptotic expansion of the reduced energy

$$\Phi(\xi) = J(v(\xi')) = V(\xi)^{p+1/p-1-N/2} c_{p,N} + O(\varepsilon). \tag{1.56}$$

To prove Theorem 3 we observe that $\frac{p+1}{p-1} - \frac{N}{2} > 0$. Then $\forall \varepsilon \ll 1$ we have

$$\inf_{\xi \in \Lambda} \Phi(\xi) < \inf_{\xi \in \partial\Lambda} \Phi(\xi) \tag{1.57}$$

and therefore Φ has a local minimum $\xi_\varepsilon \in \Lambda$ and $V(\xi_\varepsilon) \to \min_\Lambda V$. The same procedure also works for local maximums.

For several separated local minimums, the proof is similar. In fact when $|\xi_{j_1} - \xi_{j_2}| > \delta$, for all $j_1 \neq j_2$, we have $\rho = e^{-\delta_0 \min_{j_1 \neq j_2} |\xi'_{j_1} - \xi'_{j_2}|} + \varepsilon \le e^{-\delta_0 \delta/\varepsilon} + \varepsilon < 2\varepsilon$. So we obtain

$$|\nabla_x \phi(\xi')| + |\phi(\xi')| \le C\varepsilon \sum_j e^{-\delta_0 |x - \xi'_j|}. \tag{1.58}$$

Now we get

$$J(v(\xi')) = \sum_j V(\varepsilon\xi'_j)^{p+1/p-1-N/2} c_{p,N} + O(\varepsilon) \tag{1.59}$$

$\varepsilon\xi' = (\xi_1, \ldots, \xi_k)$ implies for several minimal points on the Λ_j we have the result desired.

Finally, we prove the existence of multiply interacting spikes. The computations are little bit involved since we have to measure precisely the interactions. The reduced energy functional takes the following form:

$$J(v(\xi')) = \sum_j V(\varepsilon\xi_j)^{p+1/p-1-N/2}(c_{p,N} + o(1)) - (1 + o(1))$$

$$\times \sum_{i \neq j} e^{-\min_{i \neq j}(\sqrt{V(\xi_i), V(\xi_j)})|\xi' - \xi'_j|}. \tag{1.60}$$

We shall take the following configuration space

$$\Sigma = \{(\xi_1, \ldots, \xi_k) \mid \xi_i \in \Lambda, \min_{i \neq j} |\xi_i - \xi_j| > \rho\varepsilon \log \frac{1}{\varepsilon}\}$$

and prove that the following maximization problem attains a solution in the interior part of the set Σ:

$$\min_{(\xi_1, \ldots, \xi_k) \in \Sigma} J(v(\xi')). \qquad\qquad \square$$

2. Part II: Infinite-Dimensional Reduction Method

2.1. *An introduction*

In Section 1, we have dealt with the problem of constructing solutions with finitely many bumps. The idea is to first sum up these finite many bumps and solve it in the space orthogonal to the translations. Then we adjust the points to obtain a true solution. The concentrating solutions concentrate at finite number of points which accounts for zero Lebesgue measure. In this section, we generalize this idea to the problem of constructing solutions concentrating on higher dimensional sets, such as curves, surfaces, or minimal surfaces of codimension k. As in the finite dimensional case, we proceed in two steps. In the first step, we solve the problem along each tangent fibre. This amounts to imposing *infinitely many* orthogonal conditions. In the second step, we move the higher dimensional object in the normal direction to find a true solution. We will encounter at least three problems: the first is the uniform estimate of the error in the first step. Sometimes there may be resonances due the combined effect of tangential and instability of the profile. The second problem is the adjustment of the higher dimensional subjects, which typically involves a second order nonlocal nonlinear reduced equation. The third problem is the non-compactness of the higher dimensional object.

In the following we take the model problem of Allen-Cahn equation in $\mathbb{R}^3$ and the higher dimensional concentration object is minimal surfaces. For higher dimensional concentration problems with resonances we refer to papers [19], [21] and [22].

2.2. *Model problem: The Allen-Cahn equation and minimal surfaces*

We consider the following so-called Allen-Cahn equation in $\mathbb{R}^N$

$$\Delta u + f(u) = 0 \quad \text{in } \mathbb{R}^N, \tag{2.1}$$

where $f(s) = -W'(s)$ and W is a "double-well potential", bi-stable and balanced, namely

$$W(s)>0 \text{ if } s\neq 1,-1, \quad W(1)=0=W(-1), \quad W''(\pm 1)=f'(\pm 1)=:\sigma_\pm^2>0. \tag{2.2}$$

A typical example of such a nonlinearity is

$$f(u) = (1 - u^2)u \quad \text{for } W(u) = \frac{1}{4}(1 - u^2)^2, \tag{2.3}$$

while we will not make use of the special symmetries enjoyed by this example.

Equation (2.1) is a prototype for the continuous modeling of phase transition phenomena. Let us consider the energy in a subregion region Ω of $\mathbb{R}^N$

$$J_\alpha(v) = \int_\Omega \frac{\alpha}{2} |\nabla v|^2 + \frac{1}{4\alpha} W(v),$$

whose Euler-Lagrange equation is a scaled version of (2.1),

$$\alpha^2 \Delta v + f(v) = 0 \quad \text{in } \Omega. \tag{2.4}$$

We observe that the constant functions $u = \pm 1$ minimize J_α. They are idealized as two *stable phases* of a material in Ω. It is of interest to analyze stationary configurations in which the two phases coexist. Given any subset Λ of Ω, any discontinuous function of the form

$$u_* = \chi_\Lambda - \chi_{\Omega \setminus \Lambda} \tag{2.5}$$

minimizes the second term in J_ε. The introduction of the gradient term in J_α makes an α-regularization of u_* a test function for which the energy gets bounded and proportional to the surface area of the *interface* $M = \partial\Lambda$, so that in addition to minimizing approximately the second term, stationary configurations should also select asymptotically interfaces M that are stationary for surface area, namely (generalized) minimal surfaces. This intuition on the Allen-Cahn equation gave important impulse to the calculus of variations, motivating the development of the theory of Γ-*convergence* in the 1970's. Modica [46] proved that a family of local minimizers u_α of J_α with uniformly bounded energy must converge in suitable sense to a function of the form (2.5) where $\partial\Lambda$ minimizes perimeter. Thus, intuitively, for each given $\lambda \in (-1, 1)$, the level sets $[v_\alpha = \lambda]$, collapse as $\alpha \to 0$ onto the interface $\partial\Lambda$. Similar result holds for critical points not necessarily minimizers, see [60]. For minimizers, this convergence is known in a very strong sense, see [10, 11].

If, on the other hand, we take such a critical point u_α and scale it around an interior point $0 \in \Omega$, setting $u_\alpha(x) = v_\alpha(\alpha x)$, then u_α satisfies equation (2.1) in an expanding domain,

$$\Delta u_\alpha + f(u_\alpha) = 0 \quad \text{in } \alpha^{-1}\Omega$$

so that letting formally $\alpha \to 0$ we end up with equation (2.1) in entire space. The "interface" for u_α should thus be around the (asymptotically flat) minimal surface $M_\alpha = \alpha^{-1} M$. Modica's result is based on the intuition that if M happens to be a smooth surface, then the transition from the equilibria -1 to 1 of u_α along the normal direction should take place in the approximate form $u_\alpha(x) \approx w(z)$ where z designates the normal coordinate to M_α. Then w should solve the ODE problem

$$w'' + f(w) = 0 \quad \text{in } \mathbb{R}, \quad w(-\infty) = -1, \ w(+\infty) = 1 . \qquad (2.6)$$

This solution indeed exists thanks to assumption (2.2). It is strictly increasing and unique up to constant translations. We fix in what follows the unique w for which

$$\int_\mathbb{R} t\, w'(t)^2 \, dt \ = \ 0 . \qquad (2.7)$$

For example (2.3), we have $w(t) = \tanh\left(t/\sqrt{2}\right)$. In general, w approaches its limits at exponential rates,

$$w(t) - \pm 1 \ = \ O\left(e^{-\sigma_\pm |t|}\right) \quad \text{as } t \to \pm\infty .$$

Observe then that

$$J_\alpha(u_\alpha) \approx Area\,(M) \int_\mathbb{R} [\tfrac{1}{2} w'^2 + W(w)]$$

which is what makes it plausible that M is critical for area, namely a minimal surface.

The above considerations led E. De Giorgi [24] to formulate in 1978 a celebrated conjecture on the Allen-Cahn equation (2.1), parallel to Bernstein's theorem for minimal surfaces: The level sets $[u = \lambda]$ of a bounded entire solution u to (2.1), which is also monotone in one direction, must be hyperplanes, at least for dimension $N \le 8$. Equivalently, up to a translation and a rotation, $u = w(x_1)$. This conjecture has been proven in dimensions $N = 2$ by Ghoussoub and Gui [29], $N = 3$ by Ambrosio and Cabré [9], and under a mild additional assumption by Savin [58]. A counterexample was built for $N \ge 9$ by M. del Pino, M.Kowalczyk and Wei in [25], see also [14, 43]. See [26] for a recent survey on the state of the art of this question.

The counterexample in [25] was built on the counterexample to the Bernstein conjecture for minimal graphs: Bernstein conjectured that all minimal graphs, i.e. graphs $\{x_N = F(x')\}$ for which F satisfies additionally the minimal graph equation

$$\nabla\left(\frac{\nabla F}{\sqrt{1 + |\nabla F|^2}}\right) = 0, \quad x' \in \mathbb{R}^{N-1}. \qquad (2.8)$$

In 1969, Bombierie, De Giorgi and Giusti [8] built a nontrivial solution
to (2.8) in dimension $N = 9$. In [20, 25] we took the opposite view of
Γ-convergence: for a given nondegenerate minimal surface it is possible to
build a solution to the Allen-Cahn equation which concentrates on this minimal surface. The class of minimal surfaces will include the Bombierie-De
Giorgi-Giusti minimal graph, and the complete embedded minimal surfaces
in $\mathbb{R}^3$.

In this following we construct a new class of entire solutions to the
Allen-Cahn equation in $\mathbb{R}^3$ whose level sets resemble a large dilation of a
given complete, embedded minimal surface M, asymptotically flat in the
sense that it has *finite total curvature*, namely

$$\int_M |K|\, dV \; < \; +\infty$$

where K denotes Gauss curvature of the manifold, which is also *non-
degenerate* in a sense that we will make precise below.

2.3. Embedded minimal surfaces of finite total curvature

The theory of embedded, minimal surfaces of finite total curvature in $\mathbb{R}^3$,
has reached a notable development in the last 25 years. For more than a
century, only two examples of such surfaces were known: the plane and
the catenoid. The first nontrivial example was found in 1981 by C. Costa,
[12, 13]. The *Costa surface* is a genus one minimal surface, complete and
properly embedded, which outside a large ball has exactly three components
(its *ends*), two of which are asymptotically catenoids with the same axis
and opposite directions, the third one asymptotic to a plane perpendicular
to that axis. The complete proof of embeddedness is due to Hoffman and
Meeks [34]. In [35, 37] these authors generalized notably Costa's example
by exhibiting a class of three-end, embedded minimal surface, with the
same look as Costa's far away, but with an array of tunnels that provides
arbitrary genus $k \geq 1$. This is known as the Costa-Hoffman-Meeks surface
with genus k.

Many other examples of multiple-end embedded minimal surfaces have
been found since, see for instance [44, 61] and references therein. In general
all these surfaces look like parallel planes, slightly perturbed at their ends by
asymptotically logarithmic corrections with a certain number of catenoidal

links connecting their adjacent sheets. In reality this intuitive picture is not a coincidence.

Using the Eneper-Weierstrass representation, Osserman [51] established that any embedded, complete minimal surface with finite total curvature can be described by a conformal diffeomorphism of a compact surface (actually of a Riemann surface), with a finite number of its points removed. These points correspond to the ends. Moreover, after a convenient rotation, the ends are asymptotically all either catenoids or plane, all of them with parallel axes, see Schoen [59]. The topology of the surface is thus characterized by the genus of the compact surface and the number of ends, having therefore "finite topology".

2.4. *Main results*

In what follows M designates a complete, embedded minimal surface in $\mathbb{R}^3$ with finite total curvature (to which below we will make a further nondegeneracy assumption). As pointed out in [38], M is orientable and the set $\mathbb{R}^3 \setminus M$ has exactly two components S_+, S_-.

In what follows we fix a continuous choice of unit normal field $\nu(y)$, which conventionally we take it to point towards S_+.

For $x = (x_1, x_2, x_3) = (x', x_3) \in \mathbb{R}^3$, we denote

$$r = r(x) = |(x_1, x_2)| = \sqrt{x_1^2 + x_2^2}.$$

After a suitable rotation of the coordinate axes, outside the infinite cylinder $r < R_0$ with sufficiently large radius R_0, then M decomposes into a finite number m of unbounded components $M_1, \ldots, M_m$, its *ends*. From a result in [59], we know that asymptotically each end of M_k either resembles a plane or a catenoid. More precisely, M_k can be represented as the graph of a function F_k of the first two variables,

$$M_k = \{ y \in \mathbb{R}^3 \ / \ r(y) > R_0, \ y_3 = F_k(y') \}$$

where F_k is a smooth function which can be expanded as

$$F_k(y') = a_k \log r + b_k + b_{ik} \frac{y_i}{r^2} + O(r^{-3}) \quad \text{as } r \to +\infty, \tag{2.9}$$

for certain constants a_k, b_k, b_{ik}, and this relation can also be differentiated. Here

$$a_1 \leq a_2 \leq \cdots \leq a_m, \qquad \sum_{k=1}^{m} a_k = 0. \tag{2.10}$$

The direction of the normal vector $\nu(y)$ for large $r(y)$ approaches on the ends that of the x_3 axis, with alternate signs. We use the convention that for $r(y)$ large we have

$$\nu(y) \;=\; \frac{(-1)^k}{\sqrt{1 + |\nabla F_k(y')|^2}} \left(\nabla F_k(y'), \, -1 \right) \quad \text{if } y \in M_k. \tag{2.11}$$

Let us consider the Jacobi operator of M

$$\mathcal{J}(h) \;:=\; \Delta_M h + |A|^2 h \tag{2.12}$$

where $|A|^2 = -2K$ is the Euclidean norm of the second fundamental form of M. $\mathcal{J}$ is the linearization of the mean curvature operator with respect to perturbations of M measured along its normal direction. A smooth function $z(y)$ defined on M is called a *Jacobi field* if $\mathcal{J}(z) = 0$. Rigid motions of the surface induce naturally some bounded Jacobi fields: Associated to respectively translations along coordinates axes and rotation around the x_3-axis, are the functions

$$z_1(y) = \nu(y) \cdot e_i, \quad y \in M, \quad i = 1, 2, 3,$$

$$z_4(y) = (-y_2, y_1, 0) \cdot \nu(y), \quad y \in M. \tag{2.13}$$

We assume that M is *non-degenerate* in the sense that these functions are actually *all* the bounded Jacobi fields, namely

$$\{ z \in L^\infty(M) \,/\, \mathcal{J}(z) = 0 \} \;=\; \text{span} \{ z_1, z_2, z_3, z_4 \}. \tag{2.14}$$

We denote in what follows by J the dimension (≤ 4) of the above vector space.

This assumption, expected to be generic for this class of surfaces, is known in some important cases, most notably the catenoid and the Costa-Hoffmann-Meeks surface which is an example of a three ended M whose genus may be of any order. See Nayatani [49, 50] and Morabito [47]. Note that for a catenoid, $z_{04} = 0$ so that $J = 3$. Non-degeneracy has been used as a tool to build new minimal surfaces for instance in Hauswirth and Pacard [33], and in Pérez and Ros [53]. It is also the basic element, in a compact-manifold version, to build solutions to the small-parameter Allen-Cahn equation in Pacard and Ritoré [52].

Let us consider a large dilation of M,

$$M_\alpha \;:=\; \alpha^{-1} M.$$

This dilated minimal surface has ends parameterized as

$$M_{k,\alpha} = \{\, y \in \mathbb{R}^3 \ / \ r(\alpha y) > R_0, \ y_3 = \alpha^{-1} F_k(\alpha y') \,\} \ .$$

Let β be a vector of given m real numbers with

$$\beta = (\beta_1, \ldots, \beta_m), \qquad \sum_{i=1}^{m} \beta_i = 0 \ . \tag{2.15}$$

Our main result asserts the existence of a solution $u = u_\alpha$ defined for all sufficiently small $\alpha > 0$ such that given $\lambda \in (-1, 1)$, its level set $[u_\alpha = \lambda]$ defines an embedded surface lying at a uniformly bounded distance in α from the surface M_α, for points with $r(\alpha y) = O(1)$, while its k-th end, $k = 1, \ldots, m$, lies at a uniformly bounded distance from the graph

$$r(\alpha y) > R_0, \ y_3 = \alpha^{-1} F_k(\alpha y') + \beta_k \log |\alpha y'| \ . \tag{2.16}$$

The parameters β must satisfy an additional constraint. It is clear that if two ends are parallel, say $a_{k+1} = a_k$, we need at least that $\beta_{k+1} - \beta_k \geq 0$, for otherwise the ends would eventually intersect. Our further condition on these numbers is that these ends in fact diverge at a sufficiently fast rate. We require

$$\beta_{k+1} - \beta_k > 4 \max \{\sigma_-^{-1}, \sigma_+^{-1}\} \quad \text{if} \quad a_{k+1} = a_k \ . \tag{2.17}$$

Let us consider the smooth map

$$X(y, z) = y + z\nu(\alpha y), \quad (y, t) \in M_\alpha \times \mathbb{R}. \tag{2.18}$$

$x = X(y, z)$ defines coordinates inside the image of any region where the map is one-to-one. In particular, let us consider a function $p(y)$ with

$$p(y) = (-1)^k \beta_k \log |\alpha y'| + O(1), \quad k = 1, \ldots, m,$$

and β satisfying $\beta_{k+1} - \beta_k > \gamma > 0$ for all k with $a_k = a_{k+1}$. Then the map X is one-to-one for all small α in the region of points (y, z) with

$$|z - q(y)| < \frac{\delta}{\alpha} + \gamma \log(1 + |\alpha y'|)$$

provided that $\delta > 0$ is chosen sufficiently small.

Theorem 6: *(del Pino-Kowalczyk-Wei [20]) Let $N = 3$ and M be a minimal surface embedded, complete with finite total curvature which is nondegenerate. Then, given β satisfying relations (2.15) and (2.17), there exists* a bounded solution u_α of equation (2.1), defined for all sufficiently small α, such that

$$u_\alpha(x) = w(z - q(y)) + O(\alpha) \quad \text{for all} \quad x = y + z\nu(\alpha y), \quad |z - q(y)| < \frac{\delta}{\alpha},$$
(2.19)

where the function q satisfies

$$q(y) = (-1)^k \beta_k \log |\alpha y'| + O(1) \quad y \in M_{k,\alpha}, \quad k = 1, \ldots, m.$$

In particular, for each given $\lambda \in (-1,1)$, the level set $[u_\alpha = \lambda]$ is an embedded surface that decomposes for all sufficiently small α into m disjoint components (ends) outside a bounded set. The k-th end lies at $O(1)$ distance from the graph

$$y_3 = \alpha^{-1} F_k(\alpha y) + \beta_k \log |\alpha y'|.$$

We will devote the rest of this part to the proofs of Theorems 6. For the full proofs, we refer to [20] in which more detailed behavior of the solutions constructed, such as finite Morse index, can be found.

3. Geometric Background

In this section, we present the geometric backgrounds on the expansion of the Laplacian operator near a manifold.

3.1. Parametrization of M and its Laplace-Betrami operator

Let D be the set

$$D = \{y \in \mathbb{R}^2 \, / \, |y| > R_0\}.$$

We can parameterize the end M_k of M as

$$y \in D \longmapsto y := Y_k(y) = y_i e_i + F_k(y) e_3$$
(3.1)

and F_k is the function in (2.9). In other words, for $y = (y', y_3) \in M_k$ the coodinate y is just defined as $y = y'$. We want to represent Δ_M–the Laplace-Beltrami operator of M–with respect to these coordinates. For the coefficients of the metric g_{ij} on M_k we have

$$\partial_{y_i} Y_k = e_i + O\left(r^{-1}\right) e_3$$

so that

$$g_{ij}(y) = \langle \partial_i Y_k, \partial_j Y_k \rangle = \delta_{ij} + O\left(r^{-2}\right),$$
(3.2)

where $r = |y|$. The above relations "can be differentiated" in the sense that differentiation makes the terms $O(r^{-j})$ gain corresponding negative powers of r. Then we find the representation

$$\Delta_M = \frac{1}{\sqrt{\det g_{ij}}} \partial_i(\sqrt{\det g_{ij}}\, g^{ij}\partial_j) = \Delta_{\mathbf{y}} + O(r^{-2})\partial_{ij} + O(r^{-3})\, \partial_i \quad \text{on } M_k.$$

(3.3)

The normal vector to M at $y \in M_k$ $k = 1, \ldots, m$, corresponds to

$$\nu(y) = (-1)^k \frac{1}{\sqrt{1 + |\nabla F_k(y)|^2}}\, (\partial_i F_k(y)e_i - e_3)\,, \quad y = Y_k(y) \in M_k$$

so that

$$\nu(y) = (-1)^k e_3 + \alpha_k r^{-2}\, y_i e_i + O(r^{-2})\,, \quad y = Y_k(y) \in M_k\,. \quad (3.4)$$

Let us observe for later reference that since $\partial_i \nu = O(r^{-2})$, then the principal curvatures of M, k_1, k_2 satisfy $k_l = O(r^{-2})$. In particular, we have that

$$|A(y)|^2 = k_1^2 + k_2^2 = O(r^{-4}). \quad (3.5)$$

To describe the entire manifold M we consider a finite number $N \geq m+1$ of local parametrizations

$$y \in \mathcal{U}_k \subset \mathbb{R}^2 \longmapsto y = Y_k(y), \quad Y_k \in C^\infty(\bar{\mathcal{U}}_k), \quad k = 1, \ldots, N. \quad (3.6)$$

For $k = 1, \ldots, m$, we choose them to be those in (3.1), with $\mathcal{U}_k = D$, so that $Y_k(\mathcal{U}_k) = M_k$, and $\bar{\mathcal{U}}_k$ is bounded for $k = m+1, \ldots, N$. We require then that

$$M = \bigcup_{k=1}^N Y_k(\mathcal{U}_k).$$

We remark that the Weierstrass representation of M implies that we can actually take $N = m+1$, namely only one extra parametrization is needed to describe the bounded complement of the ends in M. We will not use this fact. In general, we represent for $y \in Y_k(\mathcal{U}_k)$,

$$\Delta_M = a_{ij}^0(y)\partial_{ij} + b_i^0(y)\partial_i, \quad y = Y_k(y), \quad y \in \mathcal{U}_k, \quad (3.7)$$

where a_{ij}^0 is a uniformly elliptic matrix and the index k is not made explicit in the coefficients. For $k = 1, \ldots, m$, we have

$$a_{ij}^0(y) = \delta_{ij} + O(r^{-2}), \quad b_i^0 = O(r^{-3}), \quad \text{as } r(y) = |y| \to \infty. \quad (3.8)$$

The parametrizations set up above induce naturally a description of the expanded manifold $M_\alpha = \alpha^{-1}M$ as follows. Let us consider the functions

$$Y_{k\alpha} : \mathcal{U}_{k\alpha} := \alpha^{-1}\mathcal{U}_k \to M_\alpha, \quad y \mapsto Y_{k\alpha}(y) := \alpha^{-1}Y_k(\alpha y), \quad k = 1, \ldots, N. \tag{3.9}$$

Obviously we have

$$M_\alpha = \bigcup_{k=1}^{N} Y_{k\alpha}(\mathcal{U}_{k\alpha}).$$

The computations above lead to the following representation for the operator Δ_{M_α}:

$$\Delta_{M_\alpha} = a_{ij}^0(\alpha y)\partial_{ij} + b_i^0(\alpha y)\partial_i, \quad y = Y_{k\alpha}(y), \quad y \in \mathcal{U}_{k\alpha}, \tag{3.10}$$

where a_{ij}^0, b_i^0 are the functions in (3.7), so that for $k = 1, \ldots, m$, we have

$$a_{ij}^0 = \delta_{ij} + O(r_\alpha^{-2}), \quad b_i^0 = O(r_\alpha^{-3}), \quad \text{as } r_\alpha(y) := |\alpha y| \to \infty. \tag{3.11}$$

3.2. *Coordinates near M and the Euclidean Laplacian: Fermi coordinates*

Next we shall consider the parametrization of a neighborhood of M. Let us consider the smooth map

$$(y, z) \in M \times \mathbb{R} \longmapsto x = \tilde{X}(y, z) = y + z\nu(y) \in \mathbb{R}^3. \tag{3.12}$$

Let us consider an open subset $\tilde{\mathcal{O}}$ of $M \times \mathbb{R}$ and assume that the map $X|_{\tilde{\mathcal{O}}}$ is one to one, and that it defines a diffeomorphism onto its image $\mathcal{N} = X(\tilde{\mathcal{O}})$. Certainly we can choose $\tilde{\mathcal{O}}$ such that

$$\{(y, z) \in M \times \mathbb{R} \ / \ |z| < \delta \log(1 + r(y))\} \subset \tilde{\mathcal{O}}.$$

Since along ends $\partial_i \nu = O(r^{-2})$ so that $z\partial_i \nu$ is uniformly small in $\tilde{\mathcal{O}}$, it follows that $\tilde{X}$ is actually a diffeomorphism onto its image.

The Euclidean Laplacian Δ_x can be computed in such a region by the well-known formula in terms of the coordinates $(y, z) \in \tilde{\mathcal{O}}$ as

$$\Delta_x = \partial_{zz} + \Delta_{M_z} - H_{M_z}\partial_z, \quad x = \tilde{X}(y, z), \quad (y, z) \in \mathcal{O} \tag{3.13}$$

where M_z is the manifold

$$M_z = \{y + z\nu(y) \ / \ y \in M\}.$$

To see the formula (3.13) we observe that

$$X_i = Y_i + z\nu_i, i = 1, 2, X_z = \nu$$

and hence for $i, j = 1, 2$,

$$g_{ij}(x, z) = g_{0ij} + 2z\nu_i Y_j + z^2 \nu_i \nu_j$$

and $g_{iz} = 0, g_{zz} = 1$. Hence the Euclidean laplacian in $\tilde{\mathcal{O}}$ becomes

$$\Delta_{M_z} h(y)|_{X=X(y,z)} = \frac{1}{\sqrt{\det g_z}} \partial_i (\sqrt{\det(g_z)} g_z^{ij} \partial_i h)(y, z)$$

$$= \partial_{zz} h + \Delta_{M_z} h + \partial_z \log(\sqrt{\det g_z}) \partial_z h.$$

We note that by direct computations $\det(g_z) = \Pi_{j=1}^2 (1 - zk_j)^2 \det g_0$. This gives the formula (3.13).

Local coordinates $y = Y_k(\mathbf{y})$, $\mathbf{y} \in \mathbb{R}^2$ as in (3.1) induce natural local coordinates in M_z. The metric $g_{ij}(z)$ in M_z can then be computed as

$$g_{ij}(z) = \langle \partial_i Y, \partial_j Y \rangle + z(\langle \partial_i Y, \partial_j \nu \rangle + \langle \partial_j Y, \partial_i \nu \rangle) + z^2 \langle \partial_i \nu, \partial_j \nu \rangle \quad (3.14)$$

or

$$g_{ij}(z) = g_{ij} + z\, O(r^{-2}) + z^2 O(r^{-4})$$

where these relations can be differentiated. Thus we find from the expression of Δ_{M_z} in local coordinates that

$$\Delta_{M_z} = \Delta_M + za_{ij}^1(y, z)\partial_{ij} + zb_i^1(y, z)\partial_i, \quad y = Y(\mathbf{y}) \quad (3.15)$$

where a_{ij}^1, b_i^1 are smooth functions of their arguments. Let us examine this expansion closer around the ends of M_k where $y = Y_k(\mathbf{y})$ is chosen as in (3.1). In this case, from (3.14) and (3.2) we find

$$g^{ij}(z) = g^{ij} + z\, O(r^{-2}) + z^2 O(r^4) + \cdots.$$

Then we find that for large r,

$$\Delta_{M_z} = \Delta_M + z\, O(r^{-2})\partial_{ij} + zO(r^{-3})\partial_i. \quad (3.16)$$

Let us consider the remaining term in the expression for the Laplacian, the mean curvature H_{M_z}. We have the validity of the formula

$$H_{M_z} = \sum_{i=1}^2 \frac{k_i}{1 - k_i z} = \sum_{i=1}^2 k_i + k_i^2 z + k_i^3 z^2 + \cdots$$

where k_i, $i = 1, 2$ are the principal curvatures. Since M is a minimal surface, we have that $k_1 + k_2 = 0$. Thus

$$|A|^2 = k_1^2 + k_2^2 = -2k_1 k_2 = -2K$$

where $|A|$ is the Euclidean norm of the second fundamental form, and K the Gauss curvature. As $r \to +\infty$ we have seen that $k_i = O(r^{-2})$ and hence $|A|^2 = O(r^{-4})$. More precisely, we find for large r,

$$H_{M_z} = |A|^2 z + z^2 O(r^{-6}).$$

Thus we have found the following expansion for the Euclidean Laplacian,

$$\Delta_x = \partial_{zz} + \Delta_M - z|A|^2 \partial_z + B \tag{3.17}$$

where expressed in local coordinates in M the operator B has the form

$$B = z\, a_{ij}^1(y, z)\partial_{ij} + z\, b_i^1(y, z)\partial_i + z^2 b_3^1(y, z)\partial_z \tag{3.18}$$

with a_{ij}^1, b_i^1, b_3^1 smooth functions. Besides, we find that

$$a_{ij}^1(y, z) = O(r^{-2}), \quad b_i^1(y, z) = O(r^{-3}), \quad b_i^1(y, z) = O(r^{-6}), \tag{3.19}$$

uniformly in z for $(y, z) \in \tilde{\mathcal{O}}$. Moreover, the way these coefficients are produced from the metric yields for instance that

$$a_{ij}^1(y, z) = a_{i,j}^1(y, 0) + z a_{i,j}^{(2)}(y, z), \quad a_{i,j}^2(y, z) = O(r^{-3}),$$

$$b_i^1(y, z) = b_i^1(y, 0) + z b_i^{(2)}(y, z), \quad b_i^{(2)}(y, z) = O(r^{-4}).$$

We summarize the discussion above. Let us consider the parametrization in (3.12) of the region $\tilde{\mathcal{N}}$.

Lemma 3.1: *The Euclidean Laplacian can be expanded in $\tilde{\mathcal{N}}$ as*

$$\Delta_x = \partial_{zz} + \Delta_{M_z} - H_{M_z}\partial_z$$

$$= \partial_{zz} + \Delta_M - z|A|^2\partial_z + z\,[a_{ij}^1(y, z)\partial_{ij} + b_i^1(y, z)\partial_i] + z^2 b_3^1(y, z)\partial_z,$$

$$\Delta_M = a_{ij}^0 \partial_{ij} + b_i^0 \partial_i, \quad x = \tilde{X}(y, z), \quad (y, z) \in \tilde{\mathcal{O}},$$

where a_{ij}^l, b_j^l are smooth, bounded functions, with the index k omitted. In addition, for $k = 1, \ldots, m$,

$$a_{ij}^l = \delta_{ij}\delta_{0l} + O(r^{-2}), \quad b_i^l = O(r^{-3}), \quad b_3^l = O(r^{-6}),$$

as $r = |y| \to \infty$, uniformly in z variable.

3.3. *Laplacian in expanded variables*

Now we consider the expanded minimal surface $M_\alpha = \alpha^{-1}M$ for a small number α. We have that $\mathcal{N} = \alpha^{-1}\tilde{\mathcal{N}}$. We describe $\mathcal{N}$ via the coordinates

$$x = X(y,z) := y + z\nu_\alpha(y), \quad (y,z) \in \alpha^{-1}\tilde{\mathcal{O}}. \tag{3.20}$$

Let us observe that

$$X(y,z) = \alpha^{-1}\tilde{X}(\alpha y, \alpha z)$$

where $\tilde{x} = \tilde{X}(\tilde{y}, \tilde{z}) = \tilde{y} + \tilde{z}\nu(\tilde{y})$, where the coordinates in $\mathcal{N}_\delta$ previously dealt with. We want to compute the Euclidean Laplacian in these coordinates associated to M_α. Observe that

$$\Delta_x[u(x)]\,|_{x=X(y,z)} = \alpha^2 \Delta_{\tilde{x}}[u(\alpha^{-1}\tilde{x})]\,|_{\tilde{x}=\tilde{X}(\alpha y, \alpha z)}$$

and that the term in the right hand side is the one we have already computed. In fact setting $v(y,z) := u(y + z\nu_\alpha(y))$, we get

$$\Delta_x u\,|_{x=X(y,z)} = \alpha^2(\Delta_{\tilde{y},M_{\tilde{z}}} + \partial_{\tilde{z}\tilde{z}} - H_{M_{\tilde{z}}}\partial_{\tilde{z}})\,[v(\alpha^{-1}\tilde{y}, \alpha^{-1}\tilde{z})]\,|_{(\tilde{y},\tilde{z})=(\alpha y, \alpha z)} \cdot \tag{3.21}$$

We can then use the discussion summarized in Lemma 3.1 to obtain a representation of Δ_x in $\mathcal{N}$ via the coordinates $X(y,t)$ in (3.20). Let us consider the local coordinates $Y_{k\alpha}$ of M_α in (3.9).

Lemma 3.2: *In $\mathcal{N}$ we have*

$$\Delta_x = \partial_{zz} + \Delta_{M_{\alpha,z}} - H_{M_{\alpha,z}}\partial_z$$

$$= \partial_{zz} + \Delta_{M_\alpha} - \alpha^2 z\,|A(\alpha y)|^2\partial_z + \alpha z\,[a_{ij}^1(\alpha y, \alpha z)\partial_{ij} + \alpha b_i^1(\alpha y, \alpha z)\partial_i]$$

$$+ \alpha^3 z^2 b_3^1(\alpha y, \alpha z)\partial_z,$$

$$\Delta_{M_\alpha} = a_{ij}^0(\alpha y)\partial_{ij} + b_i^1(\alpha y)\partial_i, \quad (y,z) \in \alpha^{-1}\tilde{\mathcal{O}}, \quad y = Y_{k\alpha}(\mathbf{y})$$

where a_{ij}^l, b_j^l are smooth, bounded functions. In addition, for $k = 1, \ldots, m$,

$$a_{ij}^l = \delta_{ij}\delta_{0l} + O(r_\alpha^{-2}), \quad b_i^l = O(r_\alpha^{-3}), \quad b_3^1 = O(r_\alpha^{-6}),$$

as $r_\alpha(y) = |\alpha\mathbf{y}| \to \infty$, uniformly in z variable.

3.4. *The Euclidean Laplacian near M_α under a perturbation*

We now describe in coordinates relative to M_α the Euclidean Laplacian Δ_x, $x \in \mathbb{R}^3$, in a setting needed for the proof of our main results. The main idea is to introduce a smooth perturbation of the minimal surfaces, *a priori* unknown. We will need to compute the Euclidean Laplacian under this perturbation.

Let us consider a smooth function $h : M \to \mathbb{R}$, and the smooth map X_h defined as

$$X_h : M_\alpha \times \mathbb{R} \to \mathbb{R}^3, \quad (y,t) \longmapsto X_h(y,t) := y + (t + h(\alpha y))\,\nu(\alpha y) \quad (3.22)$$

where ν is the unit normal vector to M. Let us consider an open subset $\mathcal{O}$ of $M_\alpha \times \mathbb{R}$ and assume that the map $X_h|_{\mathcal{O}}$ is one-to-one, and that it defines a diffeomorphism onto its image $\mathcal{N} = X_h(\mathcal{O})$. Then

$$x = X_h(y,t), \quad (y,t) \in \mathcal{O},$$

defines smooth coordinates to describe the open set $\mathcal{N}$ in $\mathbb{R}^3$. Moreover, the maps

$$x = X_h(Y_{k\alpha}(\mathbf{y}),t), \quad (\mathbf{y},t) \in (\mathcal{U}_{k\alpha} \times \mathbb{R}) \cap \mathcal{O}, \quad k = 1, \ldots, N,$$

define local coordinates $(\mathbf{y},t)$ to describe the region $\mathcal{N}$. We shall assume in addition that for certain small number $\delta > 0$, we have

$$\mathcal{O} \subset \{(y,t)\ /\ |t + h(\alpha y)| < \frac{\delta}{\alpha}\,\log(2 + r_\alpha(y))\,\}. \quad (3.23)$$

We have the validity of the following expression for the Euclidean Laplacian operator in $\mathcal{N}$.

Lemma 3.3: *For $x = X_h(y,t)$, $(y,t) \in \mathcal{O}$ with $y = Y_{k\alpha}(\mathbf{y})$, $\mathbf{y} \in \mathcal{U}_{k\alpha}$, we have the validity of the identity*

$$\Delta_x = \partial_{tt} + \Delta_{M_\alpha} - \alpha^2[(t+h)|A|^2 + \Delta_M h]\partial_t - 2\alpha\,a_{ij}^0\,\partial_j h \partial_{it}$$

$$+ \alpha(t+h)\,[a_{ij}^1\partial_{ij} - 2\alpha\,a_{ij}^1\,\partial_i h \partial_{jt} + \alpha\,b_i^1\,(\partial_i - \alpha\partial_i h \partial_t)\,]$$

$$+ \alpha^3(t+h)^2 b_3^1 \partial_t + \alpha^2[a_{ij}^0 + \alpha(t+h)a_{ij}^1]\partial_i h \partial_j h\,\partial_{tt}\,. \quad (3.24)$$

Here, in agreement with (3.10), $\Delta_{M_\alpha} = a_{ij}^0(\alpha y)\partial_{ij} + b_i^0(\alpha y)\partial_i$. The functions a_{ij}^1, b_i^1, b_3^1 in the above expressions appear evaluated at the pair $(\alpha y, \alpha(t + h(\alpha y))$, while the functions h, $\partial_i h$, $\Delta_M h$, $|A|^2$, a_{ij}^0, b_i^0 are *evaluated at αy. In addition, for $k = 1, \ldots, m$, $l = 0, 1$,*

$$a_{ij}^l = \delta_{ij}\delta_{0l} + O(r_\alpha^{-2}), \quad b_i^l = O(r_\alpha^{-3}), \quad b_3^1 = O(r_\alpha^{-6}) \ ,$$

as $r_\alpha(y) = |\alpha y| \to \infty$, uniformly in their second variables. The notation $\partial_j h$ refers to $\partial_j[h \circ Y_k]$.

Proof: Let us consider a function u defined in $\mathcal{N}$, expressed in coordinates $x = X(y, z)$, and consider the expression of u in the coordinates $x = X_h(y, t)$, namely the function $v(y, t)$ defined by the relation in local coordinates $y = Y_k(y)$,

$$v(\mathbf{y}, z - h(\alpha \mathbf{y})) = u(\mathbf{y}, z),$$

(by slight abuse of notation we are denoting just by h the function $h \circ Y_k$). Then we compute

$$\partial_i u = \partial_i v - \alpha \partial_t v \partial_i h, \quad \partial_z u = \partial_t v,$$

$$\partial_{ij} u = \partial_{ij} v - \alpha \partial_{it} v \partial_j h - \alpha \partial_{jt} v \partial_i h + \alpha^2 \partial_{tt} v \partial_i h \partial_j h - \alpha^2 \partial_t v \partial_{ij} h \ .$$

Observe that, in the notation for coefficients in Lemma 3.2,

$$a_{ij}^0 \partial_{ij} h + b_i^0 \partial_i h = \Delta_M h, \quad a_{ij}^0 \partial_{ij} v + \alpha b_i^0 \partial_i v = \Delta_{M_\alpha} v \ .$$

We find then

$$\Delta_x = \partial_{tt} + \Delta_{M_\alpha} - \alpha^2 [(t + h)|A|^2 + \Delta_M h]\partial_t - 2\alpha \, a_{ij}^0 \, \partial_j h \partial_{it}$$

$$+ \alpha(t + h) \, [a_{ij}^1 \partial_{ij} - 2\alpha \, a_{ij}^1 \, \partial_i h \partial_{jt} + \alpha(b_i^1 \partial_i - \alpha b_i^1 \partial_i h \partial_t) \,]$$

$$+ \alpha^3 (t + h)^2 b_3^1 \partial_t + \alpha^2 [a_{ij}^0 + \alpha(t + h)a_{ij}^1]\partial_i h \partial_j h \, \partial_{tt} \qquad (3.25)$$

where all the coefficients are understood to be evaluated at αy or $(\alpha y, \alpha(t + h(\alpha y))$. The desired properties of the coefficients have already been established. The proof of Lemma 3.3 is concluded. $\qquad \square$

The proof actually yields that the coefficients a_{ij}^1 and b_i^1 can be further expanded as follows:

$$a_{ij}^1 = a_{ij}^1(\alpha y, 0) + \alpha(t + h) \, a_{ij}^{(2)}(\alpha y, \alpha(t + h)) =: a_{ij}^{1,0} + \alpha(t + h)a_{ij}^2,$$

with $a_{ij}^{(2)} = O(r_\alpha^{-3})$, and similarly

$$b_j^1 = b_j^1(\alpha y, 0) + \alpha(t+h)\, b_j^{(2)}(\alpha y, \alpha(t+h)) =: b_j^{1,0} + \alpha(t+h)b_j^2,$$

with $b_j^{(2)} = O(r_\alpha^{-4})$. As an example of the previous formula, let us compute the Laplacian of a function that separates variables t and y, that will be useful in Section 4.

Lemma 3.4: *Let $v(x) = k(y)\,\psi(t)$. Then the following holds.*

$$\Delta_x v = k\psi'' + \psi \Delta_{M_\alpha} k - \alpha^2[(t+h)|A|^2 + \Delta_M h]\, k\,\psi' - 2\alpha\, a_{ij}^0\, \partial_j h \partial_i k\, \psi'$$

$$+ \alpha(t+h)\,[a_{ij}^{1,0}\partial_{ij}k\psi - 2\alpha\, a_{ij}^{1,0}\, \partial_j h \partial_i k\, \psi' + \alpha(b_i^{1,0}\partial_i k\, \psi - \alpha b_i^{1,0}\partial_i h\, k\, \psi')\,]$$

$$+ \alpha^2(t+h)^2\,[a_{ij}^2\partial_{ij}k\psi - 2\alpha\, a_{ij}^2\, \partial_j h \partial_i k\, \psi' + \alpha(b_i^2\partial_i k\, \psi - \alpha b_i^2 \partial_i h\, k\, \psi')\,]$$

$$+ \alpha^3(t+h)^2 b_3^1\, k\, \psi' + \alpha^2[a_{ij}^0 + \alpha(t+h)a_{ij}^1]\partial_i h \partial_j h\, k\, \psi'' . \qquad (3.26)$$

4. Approximation of the Solution and Preliminary Discussion

4.1. *Approximation of order zero and its projection*

Let us consider a function h and sets $\mathcal{O}$ and $\mathcal{N}$ as in Section 3.4. Let $x = X_h(y,t)$ be the coordinates introduced in (3.22). At this point we shall make a more precise assumption about the function h. We need the following preliminary result whose proof we postpone for Section 7.2.

We consider a fixed m-tuple of real numbers $\beta = (\beta_1, \ldots, \beta_m)$ such that

$$\sum_{i=1}^m \beta_j = 0. \qquad (4.1)$$

Lemma 4.1: *Given any real numbers $\beta_1, \ldots, \beta_m$ satisfying (4.1), there exists a smooth function $h_0(y)$ defined on M such that*

$$J(h_0) = \Delta_M h_0 + |A|^2 h_0 = 0 \quad in\ M,$$

$$h_0(y) = (-1)^j \beta_j \log r + \theta \ \text{ as } r \to \infty \quad \text{in } M_j \quad \text{ for all } \ y \in M_j,$$

where θ satisfies

$$\|\theta\|_\infty + \|r^2 D\theta\|_\infty < +\infty . \tag{4.2}$$

We fix a function h_0 as in the above lemma and consider a function h in the form

$$h = h_0 + h_1.$$

We allow h_1 to be a parameter which we will adjust. For now we will assume that for a certain constant $\mathcal{K}$ we have

$$\|h_1\|_{L^\infty(M)} + \|(1 + r^2)Dh_1\|_{L^\infty(M)} \le \mathcal{K}\alpha. \tag{4.3}$$

We want to find a solution to

$$S(u) := \Delta_x u + f(u) = 0.$$

We consider in the region $\mathcal{N}$ the approximation

$$u_0(x) := w(t) = w(z - h_0(\alpha y) - h_1(\alpha y))$$

where z designates the normal coordinate to M_α. Thus, whenever $\beta_j \ne 0$, the level sets $[u_0 = \lambda]$ for a fixed $\lambda \in (-1, 1)$ departs logarithmically from the end $\alpha^{-1} M_j$ being still asymptotically catenoidal, more precisely it is described as the graph

$$y_3 = (\alpha^{-1} a_j + \beta_j) \log r + O(1) \text{ as } r \to \infty.$$

Note that, just as in the minimal surface case, the coefficients of the ends are balanced in the sense that they add up to zero.

It is clear that if two ends are parallel, say $a_{j+1} = a_j$, we need at least that $\beta_{j+1} - \beta_j \ge 0$, for otherwise the ends of this zero level set would eventually intersect. We recall that our further condition on these numbers is that these ends in fact diverge at a sufficiently fast rate:

$$\beta_{j+1} - \beta_j > 4 \max \{\sigma_-^{-1}, \sigma_+^{-1}\} \quad \text{if} \quad a_{j+1} = a_j . \tag{4.4}$$

We will explain later the role of this condition. Let us evaluate the error of approximation $S(u_0)$. Using Lemma 3.4 and the fact that $w'' + f(w) = 0$, we find

$$S(u_0) := \Delta_x u_0 + f(u_0)$$

$$= -\alpha^2 [|A|^2 h_1 + \Delta_M h_1] w'$$

$$- \alpha^2 |A|^2 tw' + 2 \alpha^2 a_{ij}^0 \partial_i h_0 \partial_j h_0 w''$$

$$+ \alpha^2 a_{ij}^0 (2\partial_i h_0 \partial_j h_1 + \partial_i h_1 \partial_j h_1) w''$$

$$+ 2\alpha^3 (t + h_0 + h_1) a_{ij}^1 \partial_i (h_0 + h_1) \partial_j (h_0 + h_1) w''$$

$$+ \alpha^3 (t + h_0 + h_1) b_i^1 \partial_i (h_0 + h_1) w' + \alpha^3 (t + h_0 + h_1)^3 b_3^1 w' \qquad (4.5)$$

where the formula above has been broken into "sizes", keeping in mind that h_0 is fixed while $h_1 = O(\alpha)$. Since we want that u_0 be as close as possible to be a solution of (2.1), then we would like to choose h_1 in such a way that the quantity (4.5) be as small as possible. Examining the above expression, it does not look like we can do that in absolute terms. However part of the error could be made smaller by adjusting h_1. Let us consider the "L^2-projection" onto $w'(t)$ of the error for each fixed y, given by

$$\Pi(y) := \int_{-\infty}^{\infty} S(u_0)(y, t) \, w'(t) \, dt$$

where for now, and for simplicity we assume the coordinates are defined for all t, the difference with the integration is taken in all the actual domain for t produces only exponentially small terms in α^{-1}. Then we find

$$\Pi(y) = \alpha^2 (\Delta_M h_1 + h_1 |A|^2) \int_{-\infty}^{\infty} w'^2 dt + \alpha^3 \partial_i (h_0 + h_1) \int_{-\infty}^{\infty} b_i^1 (t + h_0 + h_1) w'^2 dt$$

$$+ \alpha^3 \partial_i (h_0 + h_1) \partial_j (h_0 + h_1) \int_{-\infty}^{\infty} (t + h_0 + h) a_{ij}^1 w'' w' dt$$

$$+ \alpha^3 \int_{-\infty}^{\infty} (t + h_0 + h_1)^3 b_3^1 w'^2 dt \qquad (4.6)$$

where we have used $\int_{-\infty}^{\infty} tw'^2 \, dt = \int_{-\infty}^{\infty} w'' w' \, dt = 0$ to get rid in particular of the terms of order α^2.

Making all these "projections" equal to zero amounts to a nonlinear differential equation for h of the form

$$\mathcal{J}(h_1) = \Delta_M h_1 + h_1 |A(y)|^2 = G_0(h_1) \quad y \in M \qquad (4.7)$$

where G_0 is easily checked to be a contraction mapping of small constant in h_1, in the ball radius $O(\alpha)$ with the C^1 norm defined by the expression in the left hand side of inequality (4.3). This is where the nondegeneracy assumption on the Jacobi operator $\mathcal{J}$ enters, since we would like to invert it, in such a way to set up equation (4.7) as a fixed point problem for a contraction mapping of a ball of the form (4.3).

4.2. *Improvement of approximation*

The previous considerations are not sufficient since even after adjusting optimally h, the error in absolute value does not necessarily decrease. As we observed, the "large" term in the error,

$$-\alpha^2 |A|^2 tw' + \alpha^2 a_{ij}^0 \partial_i h_0 \partial_j h_0 \, w''$$

did not contribute to the projection. In order to eliminate, or reduce the size of this remaining part $O(\alpha^2)$ of the error, we improve the approximation through the following argument. Let us consider the differential equation

$$\psi_0''(t) + f'(w(t))\psi_0(t) = tw'(t),$$

which has a unique bounded solution with $\psi_0(0) = 0$, given explicitly by the formula

$$\psi_0(t) = w'(t) \int_0^t w'(t)^{-2} \int_{-\infty}^s sw'(s)^2 ds \,.$$

Observe that this function is well defined and it is bounded since $\int_{-\infty}^\infty sw'(s)^2 ds = 0$ and $w'(t) \sim e^{-\sigma_\pm |t|}$ as $t \to \pm\infty$, with $\sigma_\pm > 0$. Note also that $\psi_1(t) = \frac{1}{2} tw'(t)$ solves

$$\psi_1''(t) + f'(w(t))\psi_1(t) = w''(t) \,.$$

We consider as a second approximation

$$u_1 = u_0 + \phi_1, \quad \phi_1(y,t) := \alpha^2 |A(\alpha y)|^2 \psi_0(t) - \alpha^2 a_{ij}^0 \partial_i h_0 \partial_j h_0(\alpha y)\, \psi_1(t) \,. \tag{4.8}$$

Let us observe that

$$S(u_0 + \phi) = S(u_0) + \Delta_x \phi + f'(u_0)\phi + N_0(\phi), \quad N_0(\phi) = f(u_0 + \phi) - f(u_0) - f'(u_0)\phi \,.$$

We have that

$$\partial_{tt}\phi_1 + f'(u_0)\phi_1 = \alpha^2 |A(\alpha y)|^2 tw' - \alpha^2 a_{ij}^0 \partial_i h_0 \partial_j h_0(\alpha y)\, w'' \,.$$

Hence we get that the largest remaining term in the error is canceled. Indeed, we have

$$S(u_1) = S(u_0) - (2\alpha^2 a_{ij}^0 \partial_i h_0 \partial_j h_0 \, w'' - \alpha^2 |A(\alpha y)|^2 t w') + [\Delta_x - \partial_{tt}]\phi_1 + N_0(\phi_1).$$

Since ϕ_1 has size of order α^2, a smooth dependence in αy and it is of size $O(r_\alpha^{-2} e^{-\sigma|t|})$ using Lemma 3.4, we readily check that the "error created"

$$[\Delta_x - \partial_{tt}]\phi_1 + N_0(\phi_1) := -\alpha^4 \left(|A|^2 t \psi_0' - a_{ij}^0 \partial_i h_0 \partial_j h_0 \, t \psi_1' \right) \Delta h_1 + R_0$$

satisfies

$$|R_0(y,t)| \le C\alpha^3 (1 + r_\alpha(y))^{-4} e^{-\sigma|t|}.$$

Hence we have eliminated the h_1-independent term $O(\alpha^2)$ that did not contribute to the projection $\Pi(y)$, and replaced it by one smaller and with faster decay. Let us be slightly more explicit for later reference. We have

$$S(u_1) := \Delta u_1 + f(u_1)$$

$$= -\alpha^2 [|A|^2 h_1 + \Delta_M h_1] \, w' + \alpha^2 \, a_{ij}^0 \left(\partial_i h_0 \partial_j h_1 + \partial_i h_1 \partial_j h_0 + \partial_i h_1 \partial_j h_1 \right) w''$$

$$- \alpha^4 \left(|A|^2 t \psi_0' - a_{ij}^0 \partial_i h_0 \partial_j h_0 \, t \psi_1' \right) \Delta_M h_1 + 2\alpha^3 (t + h) a_{ij}^1 \, \partial_i h \partial_j h \, w'' + R_1 \tag{4.9}$$

where

$$R_1 = R_1(y, t, h_1(\alpha y), \nabla_M h_1(\alpha y))$$

with

$$|D_\imath R_1(y,t,\imath,\jmath)| + |D_\jmath R_1(y,t,\imath,\jmath)| + |R_1(y,t,\imath,\jmath)| \le C\alpha^3 (1 + r_\alpha(y))^{-4} e^{-\sigma|t|}$$

and the constant C above possibly depends on the number $\mathcal{K}$ of condition (4.3).

The above arguments are in reality the way we will actually solve the problem: two separate, but coupled steps are involved: (1) Eliminate the parts of the error that do not contribute to the projection Π and (2) Adjust h_1 so that the projection Π becomes identically zero.

4.3. *The condition of diverging ends*

Let us explain the reason to introduce condition (4.4) in the parameters β_j. To fix ideas, let us assume that we have two consecutive planar ends of M, M_j and M_{j+1}, namely with $a_j = a_{j+1}$ and with $d = b_{j+1} - b_j > 0$. Assuming that the normal in M_j points upwards, the coordinate t reads approximately as

$$t = x_3 - \alpha^{-1}b_j - h \quad \text{near } M_{j\alpha}, \quad t = \alpha^{-1}b_{j+1} - x_3 - h \quad \text{near } M_{j+1\alpha}.$$

If we let $h_0 \equiv 0$ both on $M_{j\alpha}$ and $M_{j+1\alpha}$ which are separated at distance d/α, then a good approximation in the entire region between $M_{j\alpha}$ and $M_{j+1\alpha}$ that matches the parts of $w(t)$ coming both from M_j and M_{j+1} should read near M_j approximately as

$$w(t) + w(\alpha^{-1}d - t) - 1.$$

When computing the error of approximation, we observe that the following additional term arises near $M_{j\alpha}$:

$$E := f(\,w(t) + w(\alpha^{-1}d - t) - 1\,) - f(w(t)) - f(\,w(\alpha^{-1}d - t)\,)$$

$$\sim\ [f'(w(t)) - f'(1)]\,(\,w(\alpha^{-1}d - t) - 1\,).$$

Now in the computation of the projection of the error this would give rise to

$$\int_{-\infty}^{\infty} [\,f'(w(t)) - f'(1)\,]\,(\,w(\alpha^{-1}d - t)\, - 1\,)\,w'(t)\,dt \ \sim c_* e^{-\sigma + \frac{d}{\alpha}}$$

where $c_* \neq 0$ is a constant. Thus equation (4.7) for h_1 gets modified with a term which even though very tiny, it has no decay as $|y| \to +\infty$ on M_j, unlike the others involved in the operator G_0 in (4.7). That terms eventually dominates and the equation for h_1 for very large r would read in M_j as

$$\Delta_M h_1 \sim e^{-\frac{\sigma}{\alpha}} \neq 0,$$

which is inconsistent with the assumption that h is bounded. Worse yet, its solution would be quadratic thus eventually intersecting another end. This nuisance is fixed with the introduction of h_0 satisfying condition (4.4). In that case the term E created above will now read near $M_{j\alpha}$ as

$$E \sim Ce^{-\sigma + \frac{d}{\alpha}}\,e^{-(\beta_{j+1} - \beta_j)\log r_\alpha}\,e^{-\sigma|t|} = O(e^{-\frac{\sigma}{\alpha}} r_\alpha^{-4} e^{-\sigma|t|})$$

which is qualitatively of the same type of the other terms involved in the computation of the error.

4.4. *The global first approximation*

The approximation $u_1(x)$ in (4.2) will be sufficient for our purposes, however it is so far defined only in a region of the type $\mathcal{N}$ which we have not made precise yet. Since we are assuming that M_α is connected, the fact that M_α is properly embedded implies that $\mathbb{R}^3 \setminus M_\alpha$ consists of precisely two components S_- and S_+. Let us use the convention that ν points in the direction of S_+. Let us consider the function $\mathbb{H}$ defined in $\mathbb{R}^3 \setminus M_\alpha$ as

$$\mathbb{H}(x) := \begin{cases} 1 & \text{if } x \in S_+ \\ -1 & \text{if } x \in S_- \end{cases}. \tag{4.10}$$

Then our approximation $u_1(x)$ approaches $\mathbb{H}(x)$ at an exponential rate $O(e^{-\sigma_\pm |t|})$ as $|t|$ increases. The global approximation we will use consists simply of interpolating u_1 with $\mathbb{H}$ sufficiently well-inside $\mathbb{R}^3 \setminus M_\alpha$ through a cut-off in $|t|$. In order to avoid the problem described in Section 4.3 and having the coordinates (y, t) well-defined, we consider this cut-off to be supported in a region y-dependent that expands logarithmically in r_α. Thus we will actually consider a region $\mathcal{N}_\delta$ expanding at the ends, thus becoming wider as $r_\alpha \to \infty$ than the set $\mathcal{N}_\delta^\alpha$ previously considered, where the coordinates are still well-defined.

We consider the open set $\mathcal{O}$ in $M_\alpha \times \mathbb{R}$ defined as

$$\mathcal{O} = \{(y, t) \in M_\alpha \times \mathbb{R}, \ |t + h_1(\alpha y)| < \frac{\delta}{\alpha} + 4 \max\{\sigma_-^{-1}, \sigma_+^{-1}\} \log(1 + r_\alpha(y))$$
$$=: \rho_\alpha(y)\} \tag{4.11}$$

where δ is small positive number. We consider the region $\mathcal{N} =: \mathcal{N}_\delta$ of points x of the form

$$x = X_h(y, t) = y + (t + h_0(\alpha y) + h_1(\alpha y)) \nu(\alpha y), \quad (y, t) \in \mathcal{O},$$

namely $\mathcal{N}_\delta = X_h(\mathcal{O})$. The coordinates (y, t) are well-defined in $\mathcal{N}_\delta$ for any sufficiently small δ: indeed the map X_h is one to one in $\mathcal{O}$ thanks to assumption (4.4) and the fact that $h_1 = O(\alpha)$. Moreover, Lemma 3.3 applies in $\mathcal{N}_\delta$.

Let $\eta(s)$ be a smooth cut-off function with $\eta(s) = 1$ for $s < 1$ and $= 0$ for $s > 2$. and define

$$\eta_\delta(x) := \begin{cases} \eta(|t + h_1(\alpha y)| - \rho_\alpha(y) - 3) & \text{if } x \in \mathcal{N}_\delta, \\ 0 & \text{if } x \notin \mathcal{N}_\delta \end{cases} \tag{4.12}$$

where ρ_α is defined in (4.11). Then we let our global approximation $\mathbf{w}(x)$ be simply defined as

$$\mathbf{w} := \eta_\delta u_1 + (1 - \eta_\delta)\mathbb{H} \tag{4.13}$$

where $\mathbb{H}$ is given by (4.10) and $u_1(x)$ is just understood to be $\mathbb{H}(x)$ outside $\mathcal{N}_\delta$.

Since $\mathbb{H}$ is an exact solution in $\mathbb{R}^3 \setminus M_\delta$, the global error of approximation is simply computed as

$$S(\mathbf{w}) = \Delta\mathbf{w} + f(\mathbf{w}) = \eta_\delta S(u_1) + E \tag{4.14}$$

where

$$E = 2\nabla\eta_\delta \nabla u_1 + \Delta\eta_\delta(u_1 - \mathbb{H}) + f(\eta_\delta u_1 + (1 - \eta_\delta)\mathbb{H})) - \eta_\delta f(u_1).$$

The new error terms created are of exponentially small size $O(e^{-\frac{\sigma}{\alpha}})$ but have in addition decay with r_α. In fact we have

$$|E| \leq Ce^{-\frac{\delta}{\alpha}} r_\alpha^{-4}.$$

Let us observe that $|t + h_1(\alpha y)| = |z - h_0(\alpha y)|$ where z is the normal coordinate to M_α, hence η_δ does not depend on h_1, in particular the term $\Delta\eta_\delta$ does involves second derivatives of h_1 on which we have not made assumptions yet.

5. The Proof of Theorem 6

The proof of Theorem 6 involves various ingredients whose detailed proofs are fairly technical. In order to keep the presentation as clear as possible, in this section we carry out the proof, skimming it from several (important) steps, which we state as lemmas or propositions, with complete proofs postponed for the subsequent sections.

We look for a solution u of the Allen Cahn equation (2.1) in the form

$$u = \mathbf{w} + \varphi \tag{5.1}$$

where $\mathbf{w}$ is the global approximation defined in (4.13) and φ is in some suitable sense small. Thus we need to solve the following problem

$$\Delta\varphi + f'(\mathbf{w})\varphi = -S(\mathbf{w}) - N(\varphi) \tag{5.2}$$

where

$$N(\varphi) = f(\mathbf{w} + \varphi) - f(\mathbf{w}) - f'(\mathbf{w})\varphi.$$

Next we introduce various norms that we will use to set up a suitable functional analytic scheme for solving problem (5.2). For a function $g(x)$ defined in $\mathbb{R}^3$, $1 < p \le +\infty$, $\mu > 0$, and $\alpha > 0$ we write

$$\|g\|_{p,\mu,*} := \sup_{x \in \mathbb{R}^3} (1 + r(\alpha x))^\mu \|g\|_{L^p(B(x,1))}, \quad r(x', x_3) = |x'|.$$

On the other hand, given numbers $\mu \ge 0$, $0 < \sigma < \min\{\sigma_+, \sigma_-\}$, $p > 3$, and functions $g(y, t)$ and $\phi(y, t)$ defined in $M_\alpha \times \mathbb{R}$ we consider the norms

$$\|g\|_{p,\mu,\sigma} := \sup_{(y,t) \in M_\alpha \times \mathbb{R}} r_\alpha(y)^\mu \, e^{\sigma|t|} \left(\int_{B((y,t),1)} |f|^p \, dV_\alpha \right)^{\frac{1}{p}}. \tag{5.3}$$

Consistently we set

$$\|g\|_{\infty,\mu,\sigma} := \sup_{(y,t) \in M_\alpha \times \mathbb{R}} r_\alpha(y)^\mu \, e^{\sigma|t|} \|f\|_{L^\infty(B((y,t),1))} \tag{5.4}$$

and let

$$\|\phi\|_{2,p,\mu,\sigma} := \|D^2\phi\|_{p,\mu,\sigma} + \|D\phi\|_{\infty,\mu,\sigma} + \|\phi\|_{\infty,\mu,\sigma}. \tag{5.5}$$

We consider also for a function $g(y)$ defined in M the L^p-weighted norm

$$\|f\|_{p,\beta} := \left(\int_M |f(y)|^p \, (1 + |y|^\beta)^p \, dV(y) \right)^{1/p} = \| (1 + |y|^\beta) f \|_{L^p(M)} \tag{5.6}$$

where $p > 1$ and $\beta > 0$.

We assume in what follows, that for a certain constant $\mathcal{K} > 0$ and $p > 3$ we have that the parameter function $h_1(y)$ satisfies

$$\|h_1\|_* := \|h_1\|_{L^\infty(M)} + \|(1+r^2)Dh_1\|_{L^\infty(M)} + \|D^2 h_1\|_{p,4-\frac{4}{p}} \le \mathcal{K}\alpha. \tag{5.7}$$

Next we reduce problem (5.2) to solving one qualitatively similar (equation (5.20) below) for a function $\phi(y, t)$ defined in the whole space $M_\alpha \times \mathbb{R}$.

5.1. *Step 1: The gluing reduction*

We will follow the following procedure. Let us consider again $\eta(s)$, a smooth cut-off function with $\eta(s) = 1$ for $s < 1$ and $= 0$ for $s > 2$, and define

$$\zeta_n(x) := \begin{cases} \eta(\,|t + h_1(\alpha y)| - \frac{\delta}{\alpha} + n) & \text{if } x \in \mathcal{N}_\delta \\ 0 & \text{if } x \notin \mathcal{N}_\delta \end{cases}. \tag{5.8}$$

We look for a solution $\varphi(x)$ of problem (5.2) of the following form

$$\varphi(x) = \zeta_2(x)\phi(y, t) + \psi(x) \tag{5.9}$$

where ϕ is defined in entire $M_\alpha \times \mathbb{R}$, $\psi(x)$ is defined in $\mathbb{R}^3$ and $\zeta_2(x)\phi(y,t)$ is understood as zero outside $\mathcal{N}_\delta$.

We compute, using that $\zeta_2 \cdot \zeta_1 = \zeta_1$,

$$S(\mathbf{w} + \varphi) = \Delta\varphi + f'(\mathbf{w})\varphi + N(\varphi) + S(\mathbf{w})$$

$$= \zeta_2 \left[\Delta\phi + f'(u_1)\phi + \zeta_1(f'(u_1) + H(t))\psi + \zeta_1 N(\psi + \phi) + S(u_1) \right]$$

$$+ \Delta\psi - [(1 - \zeta_1)f'(u_1) + \zeta_1 H(t)]\psi$$

$$+ (1 - \zeta_2)S(\mathbf{w}) + (1 - \zeta_1)N(\psi + \zeta_2\phi) + 2\nabla\zeta_1\nabla\phi + \phi\Delta\zeta_1 \qquad (5.10)$$

where $H(t)$ is any smooth, strictly negative function satisfying

$$H(t) = \begin{cases} f'(+1) & \text{if } t > 1 \\ f'(-1) & \text{if } t < -1 \end{cases}.$$

Thus, we will have constructed a solution $\varphi = \zeta_2\phi + \psi$ to problem (5.2) if we require that the pair (ϕ, ψ) satisfies the following coupled system

$$\Delta\phi + f'(u_1)\phi + \zeta_1(f'(u_1) - H(t))\psi + \zeta_1 N(\psi + \phi) + S(u_1) = 0 \text{ for } |t| < \frac{\delta}{\alpha} + 3 \tag{5.11}$$

$$\Delta\psi + [(1 - \zeta_1)f'(u_1) + \zeta_1 H(t)]\psi$$

$$+ (1 - \zeta_2)S(\mathbf{w}) + (1 - \zeta_1)N(\psi + \zeta_2\phi) + 2\nabla\zeta_1\nabla\phi + \phi\Delta\zeta_1 = 0 \quad \text{in } \mathbb{R}^3. \tag{5.12}$$

In order to find a solution to this system we will first extend equation (5.11) to entire $M_\alpha \times \mathbb{R}$ in the following manner. Let us set

$$B(\phi) = \zeta_4[\Delta_x - \partial_{tt} - \Delta_{y,M_\alpha}]\phi \tag{5.13}$$

where Δ_x is expressed in (y,t) coordinates using the expression (3.24) and $B(\phi)$ is understood to be zero for $|t + h_1| > \frac{\delta}{\alpha} + 5$. The other terms in equation (5.11) are simply extended as zero beyond the support of ζ_1. Thus we consider the extension of equation (5.11) given by

$$\partial_{tt}\phi + \Delta_{y,M_\alpha}\phi + \mathrm{B}(\phi) + f'(w(t))\phi = -\tilde{S}(u_1)$$

$$- \{[f'(u_1) - f'(w)]\phi + \zeta_1(f'(u_1) - H(t))\psi + \zeta_1 N(\psi + \phi)\} \text{ in } \in M_\alpha \times \mathbb{R},$$
(5.14)

where we set, with reference to expression (4.9),

$$\tilde{S}(u_1) = -\alpha^2[|A|^2 h_1 + \Delta_M h_1] w' + \alpha^2 a^0_{ij} (2\partial_i h_0 \partial_j h_1 + \partial_i h_1 \partial_j h_1) w''$$

$$- \alpha^4 (|A|^2 t\psi'_0 - a^0_{ij}\partial_i h_0 \partial_j h_0 t\psi'_1) \Delta h_1 + \zeta_4 [\alpha^3(t+h)a^1_{ij} \partial_i h \partial_j h w'' + R_1(y,t)]$$
(5.15)

and, we recall

$$R_1 = R_1(y,t,h_1(\alpha y), \nabla_M h_1(\alpha y))$$

with

$$|D_i R_1(y,t,\imath,\jmath)| + |D_j R_1(y,t,\imath,\jmath)| + |R_1(y,t,\imath,\jmath)| \leq C\alpha^3(1+r_\alpha(y))^{-4}e^{-\sigma|t|}.$$
(5.16)

In summary $\tilde{S}(u_1)$ coincides with $S(u_1)$ if $\zeta_4 = 1$ while outside the support of ζ_4, their parts that are not defined for all t are cut-off.

To solve the resulting system (5.12)-(5.14), we find first and solve equation (5.12) in ψ for a given ϕ a small function in absolute value. Noticing that the potential $[(1 - \zeta_1)f'(u_1) + \zeta_1 H(t)]$ is uniformly negative, so that the linear operator is qualitatively like $\Delta - 1$ and using contraction mapping principle, a solution $\psi = \Psi(\phi)$ is found according to the following lemma, whose detailed proof we carry out in Section 8.1.2.

Lemma 5.1: *For all sufficiently small α the following holds. Given ϕ with $\|\phi\|_{2,p,\mu,\sigma} \leq 1$, there exists a unique solution $\psi = \Psi(\phi)$ of problem (5.12) such that*

$$\|\psi\|_X := \|D^2\psi\|_{p,\mu,*} + \|\psi\|_{p,\mu,*} \leq Ce^{-\frac{\sigma\delta}{\alpha}}.$$
(5.17)

Besides, Ψ satisfies the Lipschitz condition

$$\|\Psi(\phi_1) - \Psi(\phi_2)\|_X \leq Ce^{-\frac{\sigma\delta}{\alpha}}\|\phi_1 - \phi_2\|_{2,p,\mu,\sigma}.$$
(5.18)

Thus we replace $\psi = \Psi(\phi)$ in the first equation (5.11) so that by setting

$$\mathbb{N}(\phi) := \mathbb{B}(\phi) + [f'(u_1) - f'(w)]\phi + \zeta_1(f'(u_1) - H(t))\Psi(\phi) + \zeta_1 N(\Psi(\phi) + \phi),$$
$$(5.19)$$

our problem is reduced to finding a solution ϕ to the following nonlinear, nonlocal problem in $M_\alpha \times \mathbb{R}$.

$$\partial_{tt}\phi + \Delta_{y,M_\alpha}\phi + f'(w)\phi = -\tilde{S}(u_1) - \mathbb{N}(\phi) \quad \text{in } M_\alpha \times \mathbb{R}. \qquad (5.20)$$

Thus, we concentrate in the remaining of the proof in solving equation (5.20). As we hinted in Section 4.2, we will find a solution of problem (5.20) by considering two steps: (1) "Improving the approximation", roughly solving for ϕ that eliminates the part of the error that does not contribute to the "projections" $\int[\tilde{S}(U_1) + \mathbb{N}(\phi)]w'(t)dt$, which amounts to a nonlinear problem in ϕ, and (2) Adjust h_1 in such a way that the resulting projection is actually zero. Let us set up the scheme for step (1) in a precise form.

5.2. Step 2: Eliminating terms not contributing to projections

Let us consider the problem of finding a function $\phi(y,t)$ such that for a certain function $c(y)$ defined in M_α, we have

$$\partial_{tt}\phi + \Delta_{y,M_\alpha}\phi = -\tilde{S}(u_1) - \mathbb{N}(\phi) + c(y)w'(t) \quad \text{in } M_\alpha \times \mathbb{R},$$
$$\int_{\mathbb{R}} \phi(y,t)\, w'(t)\, dt = 0, \quad \text{for all} \quad y \in M_\alpha. \qquad (5.21)$$

Solving this problem for ϕ amounts to "eliminating the part of the error that does not contribute to the projection" in problem (5.20). To justify this phrase let us consider the associated linear problem in $M_\alpha \times \mathbb{R}$

$$\partial_{tt}\phi + \Delta_{y,M_\alpha}\phi + f'(w(t))\phi = g(y,t) + c(y)w'(t), \quad \text{for all} \quad (y,t) \in M_\alpha \times \mathbb{R},$$
$$\int_{-\infty}^{\infty} \phi(y,t)\, w'(t)\, dt = 0, \quad \text{for all} \quad y \in M_\alpha. $$
$$(5.22)$$

Assuming that the corresponding operations can be carried out, let us multiply the equation by $w'(t)$ and integrate in t for fixed y. We find that

$$\Delta_{y,M_\alpha} \int_{\mathbb{R}} \phi(y,t)\, w'\, dt + \int_{\mathbb{R}} \phi(y,t)\, [w''' + f'(w)w']\, dt = \int_{\mathbb{R}} g\, w' + c(y) \int_{\mathbb{R}} w'^2.$$

The left hand side of the above identity is zero and then we find that

$$c(y) = -\frac{\int_{\mathbb{R}} g(y,t)w' dt}{\int_{\mathbb{R}} w'^2 dt},$$
(5.23)

hence a ϕ solving problem (5.22). ϕ *precisely* solves or *eliminates* the part of g which does not contribute to the projections in the equation $\Delta\phi + f'(w)\phi = g$, namely the same equation with g replaced by $\tilde{g}$ given by

$$\tilde{g}(y,t) = g(y,t) - \frac{\int_{\mathbb{R}} f(y,\cdot)w'}{\int_{\mathbb{R}} w'^2} w'(t).$$
(5.24)

The term $c(y)$ in problem (5.21) has a similar role, except that we cannot find it so explicitly.

In order to solve problem (5.21) we need to devise a theory to solve problem (5.22) where we consider a class of right hand sides g with a qualitative behavior similar to that of the error $S(u_1)$. As we have seen in (5.15), typical elements in this error are of the type $O((1+r_\alpha(y))^{-\mu}e^{-\sigma|t|})$, so this is the type of functions $g(y,t)$ that we want to consider. This is actually the motivation to introduce the norms (5.3), (5.4) and (5.5). We will prove that problem (5.22) has a unique solution ϕ which respects the size of g in norm (5.3) up to its second derivatives, namely in the norm (5.5). The following fact holds.

Proposition 5.1: *Given $p > 3$, $\mu \geq 0$ and $0 < \sigma < \min\{\sigma_-, \sigma_+\}$, there exists a constant $C > 0$ such that for all sufficiently small $\alpha > 0$ the following holds. Given f with $\|g\|_{p,\mu,\sigma} < +\infty$, then Problem (5.22) with $c(y)$ given by (5.23), has a unique solution ϕ with $\|\phi\|_{\infty,\mu,\sigma} < +\infty$. This solution satisfies in addition that*

$$\|\phi\|_{2,p,\mu,\sigma} \leq C\|g\|_{p,\mu,\sigma} .$$
(5.25)

We will prove this result in Section 6. After Proposition 5.1, solving Problem (5.21) for a small ϕ is easy using the small Lipschitz character of the terms involved in the operator $N(\phi)$ in (5.19) and contraction mapping principle. The error term $\tilde{S}(u_1)$ satisfies

$$\|\tilde{S}(u_1) + \alpha^2 \Delta h_1 w'\|_{p,4,\sigma} \leq C\alpha^3.$$
(5.26)

Using this, and the fact that $N(\phi)$ defines a contraction mapping in a ball center zero and radius $O(\alpha^3)$ in $\| \ \|_{2,p,4,\sigma}$, we conclude the existence of a unique small solution ϕ to problem (5.21) whose size is $O(\alpha^3)$ for this norm. This solution ϕ turns out to define an operator in h_1 $\phi = \Phi(h_1)$

which is Lipschitz in the norms $\| \ \|_*$ appearing in condition (5.7). In precise terms, we have the validity of the following result, whose detailed proof we postpone for Section 8.2.

Proposition 5.2: *Assume $p > 3$, $0 \le \mu \le 3$, $0 < \sigma < \min\{\sigma_+, \sigma_-\}$. There exists a $K > 0$ such that problem (8.8) has a unique solution $\phi = \Phi(h_1)$ such that*

$$\|\phi\|_{2,p,\mu,\sigma} \le K\alpha^3.$$

Besides, Φ has a Lipschitz dependence on h_1 satisfying (5.7) in the sense that

$$\|\Phi(h_1) - \Phi(h_2)\|_{2,p,\mu,\sigma} \le C\alpha^2\|h_1 - h_2\|_*. \tag{5.27}$$

5.3. Step 3: Adjusting h_1 to make the projection zero

In order to conclude the proof of the theorem, we have to carry out the second step, namely adjusting h_1, within a region of the form (5.7) for suitable $\mathcal{K}$ in such a way that the "projections" are identically zero, namely making zero the function $c(y)$ found for the solution $\phi = \Phi(h_1)$ of problem (5.21). Using expression (5.23) for $c(y)$ we find that

$$c(y)\int_{\mathbb{R}} w'^2 = \int_{\mathbb{R}} \tilde{S}(u_1)w'\,dt + \int_{\mathbb{R}} \mathbb{N}(\Phi(h_1))\,w'\,dt. \tag{5.28}$$

Now, setting $c_* := \int_{\mathbb{R}} w'^2 dt$ and using same computation employed to derive formula (4.6), we find from expression (5.15) that

$$\int_{\mathbb{R}} \tilde{S}(u_1)(y,t)\,w'(t)\,dt = -c_*\,\alpha^2(\Delta_M h_1 + h_1|A|^2) + c_*\alpha^2 G_1(h_1)$$

where

$$c_* G_1(h_1) = -\alpha^2\,\Delta h_1\,(|A|^2\int_{\mathbb{R}} t\psi_0' w'\,dt - a_{ij}^0\partial_i h_0\partial_j h_0\int_{\mathbb{R}} t\psi_1' w'\,dt)$$

$$+ \alpha\,\partial_i(h_0 + h_1)\partial_j(h_0 + h_1)\int_{\mathbb{R}} \zeta_4(t+h)a_{ij}^1 w''w'\,dt$$

$$+ \alpha^{-2}\int_{\mathbb{R}} \zeta_4\,R_1(y,t,h_1,\nabla_M h_1)w'\,dt \tag{5.29}$$

and we recall that R_1 is of size $O(\alpha^3)$ in the sense (5.16). Thus, setting

$$c_* G_2(h_1) := \alpha^{-2}\int_{\mathbb{R}} \mathbb{N}(\Phi(h_1))\,w'\,dt, \quad G(h_1) := G_1(h_1) + G_2(h_1), \tag{5.30}$$

we find that the equation $c(y) = 0$ is equivalent to the problem

$$\mathcal{J}(h_1) = \Delta_M h_1 + |A|^2 h_1 = G(h_1) \quad \text{in } M. \tag{5.31}$$

Therefore, we will have proven Theorem 6 if we find a function h_1 defined on M satisfying constraint (5.7) for a suitable $\mathcal{K}$ that solves equation (5.31). Again, this is not so direct since the operator $\mathcal{J}$ has a nontrivial bounded kernel. Rather than solving directly (5.31), we consider first a projected version of this problem, namely that of finding h_1 such that for certain scalars $c_1, \ldots, c_J$ we have

$$\mathcal{J}(h_1) = G(h_1) + \sum_{i=1}^{J} \frac{c_i}{1 + r^4} \hat{z}_i \quad \text{in } M,$$

$$\int_M \frac{\hat{z}_i h}{1 + r^4} \, dV = 0, \quad i = 1, \ldots J. \tag{5.32}$$

Here $\hat{z}_1, \ldots, \hat{z}_J$ is a basis of the vector space of bounded Jacobi fields.

In order to solve problem (5.32) we need a corresponding linear invertibility theory. This leads us to consider the linear problem

$$\mathcal{J}(h) = f + \sum_{i=1}^{J} \frac{c_i}{1 + r^4} \hat{z}_i \quad \text{in } M,$$

$$\int_M \frac{\hat{z}_i h}{1 + r^4} \, dV = 0, \quad i = 1, \ldots, J. \tag{5.33}$$

Here $\hat{z}_1, \ldots, \hat{z}_J$ are bounded, linearly independent Jacobi fields, and J is the dimension of the vector space of bounded Jacobi fields.

We will prove in Section 7.1 the following result.

Proposition 5.3: *Given $p > 2$ and f with $\|f\|_{p, 4 - \frac{4}{p}} < +\infty$, there exists a unique bounded solution h of problem (5.33). Moreover, there exists a positive number $C = C(p, M)$ such that*

$$\|h\|_* := \|h\|_\infty + \|(1 + |y|^2) Dh\|_\infty + \|D^2 h\|_{p, 4 - \frac{4}{p}} \leq C \|f\|_{p, 4 - \frac{4}{p}}. \tag{5.34}$$

Using the fact that G is a small operator of size $O(\alpha)$ uniformly on functions h_1 satisfying (5.7), Proposition 5.3 and contraction mapping principle yield the following result, whose detailed proof we carry out in Section 9.

Proposition 5.4: *Given $p > 3$, there exists a number $\mathcal{K} > 0$ such that for all sufficiently small $\alpha > 0$ there is a unique solution h_1 of problem (5.32) that satisfies constraint (5.7).*

5.4. *Step 3: Conclusion*

At the last step we prove that the constants c_i found in equation (5.32) are in reality all zero, without the need of adjusting any further parameters but rather as a consequence of the natural invariances of the of the full equation. The key point is to realize what equation has been solved so far.

First we observe the following. For each h_1 satysfying (5.7), the pair (ϕ, ψ) with $\phi = \Phi(h_1)$, $\psi = \Psi(\phi)$, solves the system

$$\Delta\phi + f'(u_1)\phi + \zeta_1(f'(u_1) - H(t))\psi + \zeta_1 N(\psi + \phi) + S(u_1)$$

$$= c(y)w'(t) \text{ for } |t| < \frac{\delta}{\alpha} + 3$$

$$\Delta\psi + [(1 - \zeta_1)f'(u_1) + \zeta_1 H(t)]\psi$$

$$+ (1 - \zeta_2)S(w) + (1 - \zeta_1)N(\psi + \zeta_2\phi) + 2\nabla\zeta_1\nabla\phi + \phi\Delta\zeta_1 = 0 \quad \text{in } \mathbb{R}^3.$$

Thus setting

$$\varphi(x) = \zeta_2(x)\phi(y, t) + \psi(x), \quad u = w + \varphi,$$

we find from formula (5.10) that

$$\Delta u + f(u) = S(w + \varphi) = \zeta_2 c(y) w'(t).$$

On the other hand choosing h_1 as that given in Proposition 5.4 which solves problem (5.32), amounts precisely to making

$$c(y) = c_*\alpha^2 \sum_{i=1}^{J} c_i \frac{\hat{z}_i(\alpha y)}{1 + r_\alpha(y)^4}$$

for certain scalars c_i. In summary, we have found h_1 satisfying constraint (5.7) such that

$$u = w + \zeta_2(x)\Phi(h_1) + \Psi(\Phi(h_1)) \tag{5.35}$$

solves the equation

$$\Delta u + f(u) = \sum_{j=1}^{J} \frac{\tilde{c}_i}{1 + r_\alpha^4} \hat{z}_i(\alpha y)w'(t) \tag{5.36}$$

where $\tilde{c}_i = c_*\alpha^2 c_i$. Testing equation (5.36) against the generators of the rigid motions $\partial_i u$ $i = 1, 2, 3$, $-x_2\partial_1 u + x_1\partial_2 u$, and using the balancing formula for the minimal surface and the zero average of the numbers β_j in the definition of h_0, we find a system of equations that leads us to $c_i = 0$ for all i, thus conclude the proof. We will carry out the details in Section 10.

In Sections 6-10, we will complete the proofs of the intermediate steps of the program designed in this section.

6. The Linearized Operator

In this section we will prove Proposition 5.1. At the core of the proof of the stated *a priori* estimates is the fact that the one-variable solution w of (2.1) is *nondegenerate* in $L^\infty(\mathbb{R}^3)$ in the sense that the linearized operator

$$L(\phi) = \Delta_y \phi + \partial_{tt}\phi + f'(w(t))\phi, \quad (y, t) \in \mathbb{R}^3 = \mathbb{R}^2 \times \mathbb{R},$$

is such that the following property holds.

Lemma 6.1: *Let ϕ be a bounded, smooth solution of the problem*

$$L(\phi) = 0 \quad in \ \mathbb{R}^2 \times \mathbb{R}. \tag{6.1}$$

Then $\phi(y, t) = Cw'(t)$ for some $C \in \mathbb{R}$.

Proof: We begin by reviewing some known facts about the one-dimensional operator $L_0(\psi) = \psi'' + f'(w)\psi$. Assuming that $\psi(t)$ and its derivative decay sufficiently fast as $|t| \to +\infty$ and defining $\psi(t) = w'(t)\rho(t)$, we get that

$$\int_\mathbb{R} [|\psi'|^2 - f'(w)\psi^2] \, dt = \int_\mathbb{R} L_0(\psi)\psi \, dt = \int_\mathbb{R} w'^2 |\rho'|^2 \, dt,$$

therefore this quadratic form is positive unless ψ is a constant multiple of w'. Using this and a standard compactness argument we get that there is a constant $\gamma > 0$ such that whenever $\int_\mathbb{R} \psi w' = 0$ with $\psi \in H^1(\mathbb{R})$ we have that

$$\int_\mathbb{R} (|\psi'|^2 - f'(w)\psi^2) \, dt \geq \gamma \int_\mathbb{R} (|\psi'|^2 + |\psi|^2) \, dt. \tag{6.2}$$

Now, let ϕ be a bounded solution of equation (6.1). We claim that ϕ has exponential decay in t, uniform in y. Let us consider a small number $\sigma > 0$ so that for a certain $t_0 > 0$ and all $|t| > t_0$ we have that

$$f'(w) < -2\sigma^2.$$

Let us consider for $\varepsilon > 0$ the function

$$g_\varepsilon(t, y) = e^{-\sigma(|t| - t_0)} + \varepsilon \sum_{i=1}^{2} \cosh(\sigma y_i).$$

Then for $|t| > t_0$ we get that

$$L(g_\delta) < 0 \quad \text{if } |t| > t_0.$$

As a conclusion, using maximum principle, we get

$$|\phi| \leq \|\phi\|_\infty \, g_\varepsilon \quad \text{if } |t| > t_0,$$

and letting $\varepsilon \to 0$ we then get

$$|\phi(y, t)| \ \leq \ C \|\phi\|_\infty e^{-\sigma|t|} \quad \text{if } |t| > t_0 .$$

Let us observe the following fact: the function

$$\tilde{\phi}(y, t) = \phi(y, t) - \left(\int_{\mathbb{R}} w'(\zeta) \, \phi(y, \zeta) \, d\zeta \right) \frac{w'(t)}{\int_{\mathbb{R}} w'^2}$$

also satisfies $L(\tilde{\phi}) = 0$ and, in addition,

$$\int_{\mathbb{R}} w'(t) \, \tilde{\phi}(y, t) \, dt = 0 \quad \text{for all} \quad y \in \mathbb{R}^2. \tag{6.3}$$

In view of the above discussion, it turns out that the function

$$\varphi(y) := \int_{\mathbb{R}} \tilde{\phi}^2(y, t) \, dt$$

is well defined. In fact so are its first and second derivatives by elliptic regularity of ϕ, and differentiation under the integral sign is thus justified. Now, let us observe that

$$\Delta_y \varphi(y) = 2 \int_{\mathbb{R}} \Delta_y \tilde{\phi} \cdot \tilde{\phi} \, dt + 2 \int_{\mathbb{R}} |\nabla_y \tilde{\phi}|^2$$

and hence

$$0 = \int_{\mathbb{R}} (L(\tilde{\phi}) \cdot \tilde{\phi})$$
$$= \frac{1}{2} \Delta_y \varphi - \int_{\mathbb{R}} |\nabla_y \tilde{\phi}|^2 \, dz - \int_{\mathbb{R}} (\,|\tilde{\phi}_t|^2 - f'(w) \tilde{\phi}^2\,) \, dt . \tag{6.4}$$

Let us observe that because of relations (6.3) and (6.2), we have that

$$\int_{\mathbb{R}} (\,|\tilde{\phi}_t|^2 - f'(w) \tilde{\phi}^2\,) \, dt \ \geq \gamma \varphi.$$

It follows then that

$$\frac{1}{2} \Delta_y \varphi - \gamma \varphi \geq 0.$$

Since φ is bounded, from maximum principle we find that φ must be identically equal to zero. But this means

$$\phi(y,t) = \left(\int_{\mathbb{R}} w'(\zeta)\, \phi(y,\zeta)\, d\zeta \right) \frac{w'(t)}{\int_{\mathbb{R}} w'^2}. \qquad (6.5)$$

Then the bounded function

$$g(y) = \int_{\mathbb{R}} w_{\zeta}(\zeta)\, \phi(y,\zeta)\, d\zeta$$

satisfies the equation

$$\Delta_y g = 0, \quad \text{in } \mathbb{R}^2. \qquad (6.6)$$

Liouville's theorem implies that $g \equiv$ constant and relation (6.5) yields $\phi(y,t) = Cw'(t)$ for some C. This concludes the proof. $\qquad \square$

6.1. A priori estimates

We shall consider problem (5.22) in a slightly more general form, also in a domain finite in y-direction. For a large number $R > 0$, let us set

$$M_{\alpha}^R := \{ y \in M_{\alpha} \ / \ r(\alpha y) < R \}$$

and consider the variation of Problem (5.22) given by

$$\partial_{tt}\phi + \Delta_{y,M_{\alpha}}\phi + f'(w(t))\phi = g(y,t) + c(y)w'(t) \quad \text{in } M_{\alpha}^R \times \mathbb{R},$$
$$\phi = 0, \quad \text{on } \partial M_{\alpha}^R \times \mathbb{R},$$
$$\int_{-\infty}^{\infty} \phi(y,t)\, w'(t)\, dt = 0 \quad \text{for all } y \in M_{\alpha}^R, \qquad (6.7)$$

where we allow $R = +\infty$ and

$$c(y) \int_{\mathbb{R}} w'^2 dt = - \int_{\mathbb{R}} g(y,t)\, w'\, dt \ .$$

We begin by proving a priori estimates.

Lemma 6.2: *Let us assume that $0 < \sigma < \min\{\sigma_-, \sigma_+\}$ and $\mu \geq 0$. Then there exists a constant $C > 0$ such that for all small α and all large R, and every solution ϕ to Problem (6.13) with $\|\phi\|_{\infty,\mu,\sigma} < +\infty$ and right hand side g satisfying $\|g\|_{p,\mu,\sigma} < +\infty$ we have*

$$\|D^2\phi\|_{p,\mu,\sigma} + \|D\phi\|_{\infty,\mu,\sigma} + \|\phi\|_{\infty,\mu,\sigma} \leq C\|g\|_{p,\mu,\sigma}. \qquad (6.8)$$

Proof: For the purpose of the *a priori* estimate, it clearly suffices to consider the case $c(y) \equiv 0$. By local elliptic estimates, it is enough to show that

$$\|\phi\|_{\infty,\mu,\sigma} \leq C\|g\|_{p,\mu,\sigma}. \tag{6.9}$$

Let us assume by contradiction that (6.9) does not hold. Then we have sequences $\alpha = \alpha_n \to 0$, $R = R_n \to \infty$, g_n with $\|g_n\|_{p,\mu,\sigma} \to 0$, ϕ_n with $\|\phi_n\|_{\infty,\mu,\sigma} = 1$ such that

$$\partial_{tt}\phi_n + \Delta_{y,M_\alpha}\phi_n + f'(w(t))\phi_n = g_n \quad \text{in } M_\alpha^R \times \mathbb{R},$$
$$\phi_n = 0 \quad \text{on } \partial M_\alpha^R \times \mathbb{R}, \tag{6.10}$$
$$\int_{-\infty}^{\infty} \phi_n(y,t)\, w'(t)\, dt = 0 \quad \text{for all } \quad y \in M_\alpha^R.$$

Then we can find points $(y_n, t_n) \in M_\alpha^R \times \mathbb{R}$ such that

$$e^{-\sigma|t_n|}(1 + r(\alpha_n y_n))^\mu\, |\phi_n(y_n, t_n)| \geq \frac{1}{2}.$$

We will consider different possibilities. We may assume that either $r_\alpha(y_n) = O(1)$ or $r_\alpha(y_n) \to +\infty$.

6.1.1. *Case $r(\alpha_n y_n)$ bounded*

We have $\alpha_n y_n$ lies within a bounded subregion of M, so we may assume that

$$\alpha_n y_n \to \tilde{y}_0 \in M.$$

Assume that $\tilde{y}_0 \in Y_k(\mathcal{U}_k)$ for one of the local parametrization of M. We consider $\tilde{y}_n, \tilde{y}_0 \in \mathcal{U}_k$ with $Y_k(\tilde{y}_n) = \alpha_n y_n$, $Y_k(\tilde{y}_0) = \tilde{y}_0$.

On $\alpha_n^{-1} Y_k(\mathcal{U}_k)$, M_α is parameterized by $Y_{k,\alpha_n}(y) = \alpha_n^{-1} Y_k(\alpha_n y)$, $y \in \alpha_n^{-1}\mathcal{U}_k$. Let us consider the local change of variable,

$$\mathbf{y} = \alpha^{-1}\tilde{\mathbf{y}}_n + \mathbf{y}.$$

6.1.2. *Subcase t_n bounded*

Let us assume first that $|t_n| \leq C$. Then, setting

$$\tilde{\phi}_n(\mathbf{y}, t) := \tilde{\phi}_n(\alpha^{-1}\tilde{\mathbf{y}}_n + \mathbf{y}, t),$$

the local equation becomes

$$a_{ij}^0(\tilde{\mathbf{y}}_n + \alpha_n \mathbf{y})\partial_{ij}\tilde{\phi}_n + \alpha_n b_j^0(\tilde{\mathbf{y}}_n + \alpha_n \mathbf{y})\partial_j\tilde{\phi}_n + \partial_{tt}\tilde{\phi}_n + f'(w(t))\tilde{\phi}_n = \tilde{g}_n(\mathbf{y}, t)$$

where $\tilde{g}_n(\mathbf{y}, t) := g_n(\tilde{\mathbf{y}}_n + \alpha \mathbf{y}, t)$. We observe that this expression is valid for y well-inside the domain $\alpha^{-1}\mathcal{U}_k$ which is expanding to entire $\mathbb{R}^2$. Since $\tilde{\phi}_n$ is bounded, and $\tilde{g}_n \to 0$ in $L_{loc}^p(\mathbb{R}^2)$, we obtain local uniform $W^{2,p}$-bound. Hence we may assume, passing to a subsequence, that $\tilde{\phi}_n$ converges uniformly in compact subsets of $\mathbb{R}^3$ to a function $\tilde{\phi}(\mathbf{y}, t)$ that satisfies

$$a_{ij}^0(\tilde{\mathbf{y}})\partial_{ij}\tilde{\phi} + \partial_{tt}\tilde{\phi} + f'(w(t))\tilde{\phi} = 0.$$

Thus $\tilde{\phi}$ is non-zero and bounded. After a rotation and stretching of coordinates, the constant coefficient operator $a_{ij}^0(\tilde{\mathbf{y}})\partial_{ij}$ becomes $\Delta_{\mathbf{y}}$. Hence Lemma 6.1 implies that, necessarily, $\tilde{\phi}(\mathbf{y}, t) = Cw'(t)$. On the other hand, we have

$$0 = \int_{\mathbb{R}} \tilde{\phi}_n(\mathbf{y}, t)\, w'(t)\, dt \longrightarrow \int_{\mathbb{R}} \tilde{\phi}(\mathbf{y}, t)\, w'(t)\, dt \quad \text{as } n \to \infty.$$

Hence, necessarily $\tilde{\phi} \equiv 0$. But we have $(1 + r(\alpha_n y_n))^\mu\, |\tilde{\phi}_n(0, t_n)| \geq \frac{1}{2}$, and since t_n and $r(\alpha_n y_n)$ were bounded, the local uniform convergence implies $\tilde{\phi} \neq 0$. We have reached a contradiction.

6.1.3. *Subcase t_n unbounded*

If y_n is in the same range as above, but, say, $t_n \to +\infty$, the situation is similar. The variation is that we define now

$$\tilde{\phi}_n(\mathbf{y}, t) = e^{\sigma(t_n + t)}\phi_n(\alpha_n^{-1}\mathbf{y}_n + \mathbf{y}, t_n + t), \quad \tilde{g}_n(\mathbf{y}, t) = e^{\sigma(t_n + t)}g_n(\alpha_n^{-1}\mathbf{y}_n + \mathbf{y}, t_n + t).$$

Then $\tilde{\phi}_n$ is uniformly bounded, and $\tilde{g}_n \to 0$ in $L_{loc}^p(\mathbb{R}^3)$. Now $\tilde{\phi}_n$ satisfies

$$a_{ij}^0(\mathbf{y}_n + \alpha_n \mathbf{y})\,\partial_{ij}\tilde{\phi}_n + \partial_{tt}\tilde{\phi}_n + \alpha_n b_j(\mathbf{y}_n + \alpha_n \mathbf{y})\,\partial_j\tilde{\phi}_n$$

$$- 2\sigma\,\partial_t\tilde{\phi}_n + (f'(w(t + t_n)) + \sigma^2)\,\tilde{\phi}_n = \tilde{g}_n.$$

We fall into the limiting situation

$$a_{ij}^*\,\partial_{ij}\tilde{\phi} + \partial_{tt}\tilde{\phi} - 2\sigma\,\partial_t\tilde{\phi} - (\sigma_+^2 - \sigma^2)\,\tilde{\phi} = 0 \quad \text{in } \mathbb{R}^3 \tag{6.11}$$

where a_{ij}^* is a positive definite, constant matrix and $\tilde{\phi} \neq 0$. But since, by hypothesis $\sigma_+^2 - \sigma^2 > 0$, maximum principle implies that $\tilde{\phi} \equiv 0$. We obtain a contradiction.

6.1.4. *Case* $r(\alpha_n y_n) \to +\infty$

In this case, we may assume that the sequence $\alpha_n y_n$ diverges along one of the ends, say M_k. Considering now the parametrization associated to the end, $y = \psi_k(\mathbf{y})$, given by (3.1), which inherits that for $M_{\alpha_n, k}$, $y = \alpha_n^{-1} \psi_k(\alpha_n \mathbf{y})$. Thus in this case $a_{ij}^0(\tilde{\mathbf{y}}_n + \alpha_n \mathbf{y}) \to \delta_{ij}$, uniformly in compact subsets of $\mathbb{R}^2$.

6.1.5. *Subcase* t_n *bounded*

Let us assume first that the sequence t_n is bounded and set
$$\tilde{\phi}_n(\mathbf{y}, t) = (1 + r(\tilde{\mathbf{y}}_n + \alpha_n \mathbf{y}))^\mu \, \phi_n(\alpha_n^{-1}\tilde{\mathbf{y}}_n + \mathbf{y}, t_n + t).$$

Then
$$\partial_j (r_{\alpha_n}^{-\mu} \tilde{\phi}_n) = -\mu \alpha \, r^{-\mu-1} \partial_j r \tilde{\phi} + r^{-\mu} \partial_j \tilde{\phi}$$

$$\partial_{ij}(r_{\alpha_n}^{-\mu}\tilde{\phi}_n) = \mu(\mu+1)\alpha^2 r^{-\mu-2}\partial_i r \partial_j r \tilde{\phi} - \mu \alpha^2 r^{-\mu-1}\partial_{ij} r \tilde{\phi} - \mu \alpha r^{-\mu-1}\partial_j r \partial_i \tilde{\phi}$$
$$+ r^{-\mu}\partial_{ij}\tilde{\phi} - \mu \alpha r^{-\mu-1}\partial_i r \partial_j \tilde{\phi} \ .$$

Now $\partial_i r = O(1)$, $\partial_{ij} r = O(r^{-1})$, hence we have
$$\partial_j(r_{\alpha_n}^{-\mu}\tilde{\phi}_n) = r^{-\mu}\left[\partial_j \tilde{\phi} + O(\alpha r_\alpha^{-1})\tilde{\phi}\right] \, ,$$

$$\partial_{ij}(r_{\alpha_n}^{-\mu}\tilde{\phi}_n) = r_\alpha^{-\mu}\left[\partial_{ij}\tilde{\phi} + O(\alpha r_\alpha^{-1})\partial_i \tilde{\phi} + O(\alpha^2 r_\alpha^{-2})\tilde{\phi}\right] \, ,$$

and the equation satisfied by $\tilde{\phi}_n$ has therefore the form
$$\Delta_{\mathbf{y}}\tilde{\phi}_n + \partial_{tt}\tilde{\phi}_n + o(1)\partial_{ij}\tilde{\phi}_n + o(1)\,\partial_j \tilde{\phi}_n + o(1)\,\tilde{\phi}_n + f'(w(t))\tilde{\phi}_n = \tilde{g}_n$$
where $\tilde{\phi}_n$ is bounded, $\tilde{g}_n \to 0$ in $L^p_{loc}(\mathbb{R}^3)$. From elliptic estimates, we also get uniform bounds for $\|\partial_j \tilde{\phi}_n\|_\infty$ and $\|\partial_{ij}\tilde{\phi}_n\|_{p,0,0}$. In the limit we obtain a $\tilde{\phi} \neq 0$ bounded, solution of
$$\Delta_{\mathbf{y}}\tilde{\phi} + \partial_{tt}\tilde{\phi} + f'(w(t))\tilde{\phi} = 0, \quad \int_{\mathbb{R}}\tilde{\phi}(\mathbf{y}, t)\, w'(t)\, dt \ = \ 0, \qquad (6.12)$$

a situation which is discarded in the same way as before if $\tilde{\phi}$ is defined in $\mathbb{R}^3$. There is however, one more possibility which is that $r(\alpha_n y_n) - R_n = O(1)$. In such a case we would see in the limit equation (6.12) satisfied in a half-space, which after a rotation in the $\mathbf{y}$-plane can be assumed to be

$H = \{(\mathbf{y}, t) \in \mathbb{R}^2 \times \mathbb{R} \, / \, y_2 < 0\}, \quad \text{with } \phi(y_1, 0, t) = 0 \quad \text{for all} \quad (y_1, t) \in \mathbb{R}^2.$

By Schwarz's reflection, the odd extension of $\tilde{\phi}$, which achieves for $y_2 > 0$, $\tilde{\phi}(y_1, y_2, t) = -\tilde{\phi}(y_1, -y_2, t)$, satisfies the same equation, and thus we fall into one of the previous cases, again finding a contradiction.

6.1.6. *Subcase t_n unbounded*

Let us assume now $|t_n| \to +\infty$. If $t_n \to +\infty$ we define

$$\tilde{\phi}_n(\mathbf{y}, t) = (1 + r(\tilde{\mathbf{y}}_n + \alpha_n \mathbf{y}))^\mu \, e^{t_n + t} \, \phi_n(\alpha_n^{-1} \tilde{\mathbf{y}}_n + \mathbf{y}, t_n + t).$$

In this case, we end up in the limit with a $\tilde{\phi} \neq 0$ bounded and satisfying the equation

$$\Delta_{\mathbf{y}} \tilde{\phi} + \partial_{tt} \tilde{\phi} - 2\sigma \, \partial_t \tilde{\phi} - (\sigma_+^2 - \sigma^2) \, \tilde{\phi} = 0$$

either in entire space or in a half-space under zero boundary condition. This implies again $\tilde{\phi} = 0$, and a contradiction has been reached that finishes the proof of the *a priori* estimates.

6.2. **Existence: Conclusion of proof of Proposition 5.1**

Let us prove now existence. We assume first that g has compact support in $M_\alpha \times \mathbb{R}$.

$$\partial_{tt} \phi + \Delta_{y, M_\alpha} \phi + f'(w(t))\phi = g(y, t) + c(y)w'(t) \quad \text{in } M_\alpha^R \times \mathbb{R},$$
$$\phi = 0, \quad \text{on } \partial M_\alpha^R \times \mathbb{R},$$
$$\int_{-\infty}^{\infty} \phi(y, t) \, w'(t) \, dt = 0 \quad \text{for all} \quad y \in M_\alpha^R, \qquad (6.13)$$

where we allow $R = +\infty$ and

$$c(y) \int_{\mathbb{R}} w'^2 dt = - \int_{\mathbb{R}} g(y, t) \, w' \, dt \ .$$

Problem (6.13) has a weak formulation which is the following. Let

$$H = \{ \phi \in H_0^1(M_\alpha^R \times \mathbb{R}) \, / \int_{\mathbb{R}} \phi(y, t) \, w'(t) \, dt = 0 \quad \text{for all} \quad y \in M_\alpha^R \} \, .$$

H is a closed subspace of $H_0^1(M_\alpha^R \times \mathbb{R})$, hence a Hilbert space when endowed with its natural norm,

$$\|\phi\|_H^2 = \int_{M_\alpha^R} \int_{\mathbb{R}} \left(|\partial_t \phi|^2 + |\nabla_{M_\alpha} \phi|^2 - f'(w(t)) \, \phi^2 \right) dV_\alpha \, dt \ .$$

ϕ is then a weak solution of Problem (6.13) if $\phi \in H$ and satisfies

$$a(\phi, \psi) := \int_{M_\alpha^R \times \mathbb{R}} \left(\nabla_{M_\alpha} \phi \cdot \nabla_{M_\alpha} \psi - f'(w(t)) \, \phi \psi \right) dV_\alpha \, dt$$
$$= - \int_{M_\alpha^R \times \mathbb{R}} g \, \psi \, dV_\alpha \, dt \quad \text{for all} \quad \psi \in H.$$

It is standard to check that a weak solution of problem (6.13) is also classical provided that g is regular enough. Let us observe that because of the orthogonality condition defining H we have that

$$\gamma \int_{M_\alpha^R \times \mathbb{R}} \psi^2 \, dV_\alpha \, dt \leq a(\psi, \psi) \quad \text{for all} \quad \psi \in H.$$

Hence the bilinear form a is coercive in H, and existence of a unique weak solution follows from Riesz's theorem. If g is regular and compactly supported, ψ is also regular. Local elliptic regularity implies in particular that ϕ is bounded. Since for some $t_0 > 0$, the equation satisfied by ϕ is

$$\Delta \phi + f'(w(t)) \phi = c(y) w'(t), \quad |t| > t_0, \quad y \in M_\alpha^R, \tag{6.14}$$

and $c(y)$ is bounded, then enlarging t_0 if necessary, we see that for $\sigma < \min\{\sigma_+, \sigma_-\}$, the function $v(y, t) := Ce^{-\sigma|t|} + \varepsilon e^{\sigma|t|}$ is a positive supersolution of equation (6.14), for a large enough choice of C and arbitrary $\varepsilon > 0$. Hence $|\phi| \leq Ce^{-\sigma|t|}$, from maximum principle. Since M_α^R is bounded, we conclude that $\|\phi\|_{p,\mu,\sigma} < +\infty$. From Lemma 6.2 we obtain that if R is large enough then

$$\|D^2\phi\|_{p,\mu,\sigma} + \|D\phi\|_{\infty,\mu,\sigma} + \|\phi\|_{\infty,\mu,\sigma} \leq C\|g\|_{p,\mu,\sigma}. \tag{6.15}$$

Now let us consider Problem (6.13) for $R = +\infty$, allowed above, and for $\|g\|_{p,\mu,\sigma} < +\infty$. Then solving the equation for finite R and suitable compactly supported g_R, we generate a sequence of approximations ϕ_R which is uniformly controlled in R by the above estimate. If g_R is chosen so that $g_R \to g$ in $L^p_{loc}(M_\alpha \times \mathbb{R})$ and $\|g_R\|_{p,\mu,\sigma} \leq C\|g\|_{p,\mu,\sigma}$, we obtain that ϕ_R is locally uniformly bounded, and by extracting a subsequence, it converges uniformly locally over compacts to a solution ϕ to the full problem which respects the estimate (5.25). This concludes the proof of existence, and hence that of the proposition.

7. Theory of the Jacobi Operator

We consider this section the problem of finding a function h such that for certain constants $c_1, \ldots, c_J$,

$$\mathcal{J}(h) = \Delta_M h + |A|^2 h = f + \sum_{j=1}^{J} \frac{c_i}{1 + r^4} \, \hat{z}_i \quad \text{in } M, \tag{7.1}$$

$$\int_M \frac{\hat{z}_i h}{1 + r^4} = 0, \quad i = 1, \ldots, J \tag{7.2}$$

and prove the result of Proposition 5.3. We will also deduce the existence of Jacobi fields of logarithmic growth as in Lemma 4.1. We recall the definition of the norms $\| \ \|_{p,\beta}$ in (5.6).

Outside of a ball of sufficiently large radius R_0, it is natural to parameterize each end of M, $y_3 = F_k(y_1, y_2)$ using the Euclidean coordinates $\mathbf{y} = (y_1, y_2) \in \mathbb{R}^2$. The requirement in f on each end amounts to $\tilde{f} \in L^p(B(0, 1/R_0))$ where

$$\tilde{f}(\mathbf{y}) := |\mathbf{y}|^{-4} f(|\mathbf{y}|^{-2}\mathbf{y}) \,. \tag{7.3}$$

Indeed, observe that

$$\|\tilde{f}\|^p_{L^p(B(0,1/R_0))} = \int_{B(0,1/R_0)} |\mathbf{y}|^{-4p} |f(|\mathbf{y}|^{-2}\mathbf{y})|^p \, d\mathbf{y}$$

$$= \int_{\mathbb{R}^2 \setminus B(0,R_0)} |\mathbf{y}|^{4(p-1)} |f(\mathbf{y})|^p \, d\mathbf{y} \,.$$

In order to prove the proposition we need some *a priori* estimates.

Lemma 7.1: *Let $p > 2$. For each $R_0 > 0$ sufficiently large there exists a constant $C > 0$ such that if*

$$\|f\|_{p,4-\frac{4}{p}} + \|h\|_{L^\infty(M)} < +\infty$$

and h solves

$$\Delta_M h + |A|^2 h = f, \quad y \in M, \quad |y| > R_0 \,,$$

then

$$\|h\|_{L^\infty(|y|>2R_0)} + \| \, |y|^2 Dh\|_{L^\infty(|y|>2R_0)} + \| \, |y|^{4-\frac{4}{p}} D^2 h\|_{L^p(|y|>2R_0)}$$

$$\leq C \left[\|f\|_{p,4-\frac{4}{p}} + \|h\|_{L^\infty(R_0<|y|<3R_0)} \right] \,.$$

Proof: Along each end M_k of M, Δ_M can be expanded in the coordinate $\mathbf{y}$ as

$$\Delta_M = \Delta + O(|\mathbf{y}|^{-2}) D^2 + O(|\mathbf{y}|^{-3}) D.$$

A solution of h of equation (7.1) satisfies

$$\Delta_M h + |A|^2 h = f, \quad |\mathbf{y}| > R_0$$

for a sufficiently large R_0. Let us consider a Kelvin's transform

$$h(\mathbf{y}) = \tilde{h}(\mathbf{y}/|\mathbf{y}|^2).$$

Then we get

$$\Delta h(\mathbf{y}) = |\mathbf{y}|^{-4} (\Delta \tilde{h})(\mathbf{y}/|\mathbf{y}|^2) \,.$$

Besides

$$O(|\mathbf{y}|^{-2})D^2h(\mathbf{y})+O(|\mathbf{y}|^{-3})Dh(\mathbf{y})=O(|\mathbf{y}|^{-6})D^2\tilde{h}(\mathbf{y}/|\mathbf{y}|^2)+O(|\mathbf{y}|^{-5})D\tilde{h}(\mathbf{y}/|\mathbf{y}|^2).$$

Hence

$$(\Delta_M h)(\mathbf{y}/|\mathbf{y}|^2) = |\mathbf{y}|^4 \left[\Delta\tilde{h}(\mathbf{y}) + O(|\mathbf{y}|^2)D^2\tilde{h}(\mathbf{y}) + O(|\mathbf{y}|)D\tilde{h}(\mathbf{y}) \right].$$

Then $\tilde{h}$ satisfies the equation

$$\Delta\tilde{h} + O(|\mathbf{y}|^2)D^2\tilde{h} + O(|\mathbf{y}|)D\tilde{h} + O(1)h = \tilde{f}(\mathbf{y}), \quad 0 < |\mathbf{y}| < \frac{1}{R_0}$$

where $\tilde{f}$ is given by (7.3). The operator above satisfies maximum principle in $B(0,\frac{1}{R_0})$ if R_0 is fixed large enough. This, the fact that $\tilde{h}$ is bounded, and L^p-elliptic regularity for $p > 2$ in two dimensional space imply that

$$\|\tilde{h}\|_{L^\infty(B(0,1/2R_0))} + \|D\tilde{h}\|_{L^\infty(B(0,1/2R_0))} + \|D^2\tilde{h}\|_{L^p(B(0,1/2R_0))}$$

$$\leq C[\|\tilde{f}\|_{L^p((B(0,1/R_0))} + \|\tilde{h}\|_{L^\infty(1/3R_0<|\mathbf{y}|<1/R_0)}]$$

$$\leq C\left[\|f\|_{p,4-\frac{4}{p}} + \|h\|_{L^\infty(B(R_0<|\mathbf{y}|<3R_0))}\right].$$

Let us observe that

$$\|\tilde{h}\|_{L^\infty(B(0,1/2R_0))} = \|h\|_{L^\infty(|\mathbf{y}|>2R_0)},$$

$$\|D\tilde{h}\|_{L^\infty(B(0,1/2R_0))} = \| |\mathbf{y}|^2 Dh\|_{L^\infty(|\mathbf{y}|>2R_0)}.$$

Since

$$|D^2h(\mathbf{y})| \leq C(|\mathbf{y}|^{-4}|D^2\tilde{h}(|\mathbf{y}|^{-2}\mathbf{y})| + |\mathbf{y}|^{-3}|D\tilde{h}(|\mathbf{y}|^{-2}\mathbf{y})|)$$

then

$$|\mathbf{y}|^{4-\frac{4}{p}}|D^2h(\mathbf{y})| \leq C(|\mathbf{y}|^{-4/p}|D^2\tilde{h}(|\mathbf{y}|^{-2}\mathbf{y})| + |\mathbf{y}|^{-\frac{4}{p}-1}|D\tilde{h}(|\mathbf{y}|^{-2}\mathbf{y})|).$$

Hence

$$\int_{|\mathbf{y}|>2R_0} |\mathbf{y}|^{4p-4}|D^2h|^p d\mathbf{y}$$

$$\leq C(\int_{B(0,1/2R_0)} |D^2\tilde{h}(\mathbf{y})|^p \, d\mathbf{y} + \|D\tilde{h}\|_{L^\infty(B(0,1/2R_0))}^p \int_{|\mathbf{y}|>2R_0} |\mathbf{y}|^{-4-p} d\mathbf{y}).$$

It follows that

$$\|h\|_{L^\infty(|\mathbf{y}|>2R_0)} + \| |\mathbf{y}|^2 Dh\|_{L^\infty(|\mathbf{y}|>2R_0)} + \| |\mathbf{y}|^{4-\frac{4}{p}}D^2h\|_{L^p(|\mathbf{y}|>2R_0)}$$

$$\leq C \left[\|f\|_{p,4-\frac{4}{p}} + \|h\|_{L^\infty(B(R_0 < |y| < 3R_0))} \right].$$

Since this estimate holds at each end, the result of the lemma follows, after possibly changing slightly the value R_0. □

Lemma 7.2: *Under the conditions of Lemma 7.1, assume that h is a bounded solution of Problem (7.1)-(7.2). Then the a priori estimate (5.34) holds.*

Proof: Let us observe that this *a priori* estimate in Lemma 7.1 implies in particular that the Jacobi fields $\hat{z}_i$ satisfy

$$\nabla \hat{z}_i(y) = O(|y|^{-2}) \quad \text{as } |y| \to +\infty.$$

Using $\hat{z}_i$ as a test function in a ball $B(0,\rho)$ in M we obtain

$$\int_{\partial B(0,\rho)} (h \partial_\nu \hat{z}_i - \hat{z}_i \partial_\nu \hat{z}_i) + \int_{|y|<\rho} (\Delta_M \hat{z}_i + |A|^2 \hat{z}_i) h$$

$$= \int_{|y|<\rho} f \hat{z}_i + \sum_{j=1}^J c_j \int_M \frac{\hat{z}_i \hat{z}_j}{1+r^4}.$$

Since the boundary integral in the above identity is of size $O(\rho^{-1})$ we get

$$\int_M f \hat{z}_i + \sum_{j=1}^J c_j \int_M \frac{\hat{z}_i \hat{z}_j}{1+r^4} = 0 \tag{7.4}$$

so that in particular

$$|c_j| \leq C \|f\|_{p,4-\frac{4}{p}} \quad \text{for all} \quad j = 1,\ldots,J. \tag{7.5}$$

In order to prove the desired estimate, we assume by contradiction that there are sequences h_n, f_n with $\|h_n\|_\infty = 1$ and $\|f_n\|_{p,4-\frac{4}{p}} \to 0$, such that

$$\Delta_M h_n + |A|^2 h_n = f_n + \sum_{j=1}^J \frac{c_i^n \hat{z}_i}{1+r^4}$$

$$\int_M \frac{h_n \hat{z}_i}{1+r^4} = 0 \quad \text{for all} \quad i = 1,\ldots,J.$$

Thus according estimate (7.5), we have that $c_i^n \to 0$. From Lemma 7.1 we find

$$\|h_n\|_{L^\infty(|y|>2R_0)} \le C[o(1) + \|h_n\|_{L^\infty(B(0,3R_0))}].$$

The latter inequality implies that

$$\|h_n\|_{L^\infty(B(0,3R_0))} \ge \gamma > 0.$$

Local elliptic estimates imply a C^1 bound for h_n on bounded sets. This implies the presence of a subsequence h_n which we denote the same way such that $h_n \to h$ uniformly on compact subsets of M, where h satisfies

$$\Delta_M h + |A|^2 h = 0.$$

h is bounded hence, by the nondegeneracy assumption, it is a linear combination of the functions $\hat{z}_i$. Besides $h \ne 0$ and satisfies

$$\int_M \frac{h\hat{z}_i}{1+r^4} = 0 \quad \text{for all} \quad i = 1, \ldots, J.$$

The latter relations imply $h = 0$, hence a contradiction that proves the validity of the *a priori* estimate. □

7.1. *Proof of Proposition 5.3*

Thanks to Lemma 7.2 it only remains to prove existence of a bounded solution to Problems (7.1)-(7.2). Let f be as in the statement of the proposition. Let us consider the Hilbert space H of functions $h \in H^1_{loc}(M)$ with

$$\|h\|_H^2 := \int_M |\nabla h|^2 + \frac{1}{1+r^4}|h|^2 < +\infty,$$

$$\int_M \frac{1}{1+r^4} h\hat{z}_i = 0 \quad \text{for all} \quad i = 1, \ldots, J.$$

Problems (7.1)-(7.2) can be formulated in weak form as that of finding $h \in H$ with

$$\int_M \nabla h \nabla \psi - |A|^2 h \psi = -\int_M f\psi \quad \text{for all} \quad \psi \in H.$$

In fact, a weak solution $h \in H$ of this problem must be bounded thanks to elliptic regularity, with the use of Kelvin's transform in each end for the control at infinity. Using that $|A|^2 \le Cr^{-4}$, Riesz representation theorem and the fact that H is compactly embedded in $L^2((1+r^4)^{-1}dV)$ (which follows for instance by inversion at each end), we see that this weak problem can be written as an equation of the form

$$h - T(h) = \tilde{f}$$

where T is a compact operator in H and $\tilde{f} \in H$ depends linearly on f. When $f = 0$, the *a priori* estimates found yield that necessarily $h = 0$. Existence of a solution then follows from Fredholm's alternative. The proof is complete.

7.2. Jacobi fields of logarithmic growth. The proof of Lemma 4.1

We will use the theory developed above to construct Jacobi fields with logarithmic growth as $r \to +\infty$, whose existence we stated and use to set up the initial approximation in Lemma 4.1. One of these Jacobi fields is the generator of dilations of the surface, $z_0(y) = y \cdot \nu(y)$. We will prove next that there are another $m - 2$ linearly independent logarithmically growing Jacobi fields.

Let us consider an m-tuple of numbers $\beta_1, \ldots, \beta_m$ with $\sum_j \beta_j = 0$, and any smooth function $p(y)$ in M such that on each end M_j we have that for sufficiently large $r = r(y)$,

$$p(y) = (-1)^j \beta_j \log r(y), \quad y \in M_j$$

for certain numbers $\beta_1, \ldots, \beta_m$ that we will choose later. To prove the result of Lemma 4.1 we need to find a solution h_0 of the equation $\mathcal{J}(h_0) = 0$ of the form $h_0 = p + h$ where h is bounded. This amounts to solving

$$\mathcal{J}(h) = -\mathcal{J}(p). \tag{7.6}$$

Let us consider the cylinder $C_R = \{x \in \mathbb{R}^3 \, / \, r(x) < R\}$ for a large R. Then

$$\int_{M \cap C_R} \mathcal{J}(p) \, z_3 dV = \int_{M \cap C_R} \mathcal{J}(z_3) z_3 dV + \int_{\partial C_R \cap M} (z_3 \partial_n p - p \partial_n z_3) \, d\sigma(y).$$

Hence

$$\int_{M \cap C_R} \mathcal{J}(p) \, z_3 dV = \sum_{j=1}^m \int_{\partial C_R \cap M_j} (z_3 \partial_n p - p \partial_n z_3) \, d\sigma(y).$$

Thus using the graph coordinates on each end, we find

$$\int_{M \cap C_R} \mathcal{J}(p) \, z_3 dV$$

$$= \sum_{j=1}^m (-1)^j \left[\frac{\beta_j}{R} \int_{|y|=R} \nu_3 d\sigma(\mathbf{y}) - \beta_j \log R \int_{|y|=R} \partial_r \nu_3 d\sigma(\mathbf{y}) \right] + O(R^{-1}).$$

We have that, on each end M_j,

$$\nu_3(\mathbf{y}) = \frac{(-1)^j}{\sqrt{1 + |\nabla F_k(\mathbf{y})|^2}} = (-1)^j + O(r^{-2}), \quad \partial_r \nu_3(\mathbf{y}) = O(r^{-3}).$$

Hence we get

$$\int_{M \cap C_R} \mathcal{J}(p) z_3 dV = 2\pi \sum_{j=1}^m \beta_j + O(R^{-1}).$$

It is easy to see, using the graph coordinates that $\mathcal{J}(p) = O(r^{-4})$ and it is hence integrable. We pass to the limit $R \to +\infty$ and get

$$\int_M \mathcal{J}(p) z_3 dV = 2\pi \sum_{j=1}^m \beta_j = 0 . \qquad (7.7)$$

We make a similar integration for the remaining bounded Jacobi fields. For $z_i = \nu_i(y)$ $i = 1, 2$ we find

$$\int_{M \cap C_R} \mathcal{J}(p) z_2 dV$$

$$= \sum_{j=1}^m (-1)^j \left[\frac{\beta_j}{R} \int_{|y|=R} \nu_2 d\sigma(\mathbf{y}) - \beta_j \log R \int_{|y|=R} \partial_r \nu_2 d\sigma(\mathbf{y}) \right] + O(R^{-1}).$$

Now, on M_j,

$$\nu_2(\mathbf{y}) = \frac{(-1)^j}{\sqrt{1 + |\nabla F_k(\mathbf{y})|^2}} = (-1)^j a_j \frac{x_i}{r^2} + O(r^{-3}), \quad \partial_r \nu_2(\mathbf{y}) = O(r^{-2}).$$

Hence

$$\int_M \mathcal{J}(p) z_i dV = 0 \ i = 1, 2.$$

Finally, for $z_4(y) = (-y_2, y_1, 0) \cdot \nu(y)$ we find on M_j,

$$(-1)^j z_4(\mathbf{y}) = -\mathbf{y}_2 \partial_2 F_j + \mathbf{y}_1 \partial_1 F_j = b_{j1} \frac{\mathbf{y}_2}{r^2} - b_{j2} \frac{\mathbf{y}_1}{r^2} + O(r^{-2}), \quad \partial_r z_4 = O(r^{-2})$$

and hence again

$$\int_M \mathcal{J}(p) z_4 dV = 0 .$$

From the solvability theory developed, we can then find a bounded solution to the problem

$$\mathcal{J}(h) = -\mathcal{J}(p) + \sum_{j=1}^J q c_j \hat{z}_j .$$

Since $\int_M \mathcal{J}(p) z_i dV = 0$ and hence $\int_M \mathcal{J}(p) \hat{z}_i dV = 0$, relations (7.4) imply that $c_i = 0$ for all i.

We have thus found a bounded solution to equation (7.6) and the proof is concluded.

Remark 7.3: *Observe that, in particular, the explicit Jacobi field* $z_0(y) = y \cdot \nu(y)$ *satisfies that*

$$z(y) = (-1)^j a_j \log r + O(1) \quad for \ all \quad y \in M_j$$

and we have indeed $\sum_j a_j = 0$. *Besides this one, we thus have the presence of another* $m - 2$ *linearly independent Jacobi fields with* $|z(y)| \sim \log r$ *as* $r \to +\infty$, *where* m *is the number of ends.*

These are in reality all Jacobi fields with exact logarithmic growth. In fact if $\mathcal{J}(z) = 0$ *and*

$$|z(y)| \le C \log r , \qquad (7.8)$$

then the argument in the proof of Lemma 7.1 shows that the Kelvin's inversion $\tilde{z}(\mathbf{y})$ *as in the proof of Lemma 7.2 satisfies near the origin* $\Delta \tilde{z} = \tilde{f}$ *where* $\tilde{f}$ *belongs to any* L^p *near the origin, so it must equal a multiple of* $\log |\mathbf{y}|$ *plus a regular function. It follows that on* M_j *there is a number* β_j *with*

$$z(\mathbf{y}) = (-1)^j \beta_j \log |\mathbf{y}| + h$$

where h *is smooth and bounded. The computations above force* $\sum_j \beta_j = 0$. *It follows from Lemma 4.1 that then* z *must be equal to one of the elements there predicted plus a bounded Jacobi field. We conclude in particular that the dimension of the space of Jacobi fields satisfying (7.8) must be at most* $m - 1 + J$, *thus recovering a fact stated in Lemma 5.2 of* [53].

8. Reducing the Gluing System and Solving the Projected Problem

In this section, we prove Lemma 5.1, which reduces the gluing system (5.12)-(5.14) to solving the nonlocal equation (5.20) and prove Proposition 5.2 on solving the nonlinear projected problem (5.21), in which the basic element is linear theory stated in Proposition 5.1. In what follows we refer to notation and objects introduced in Sections 5.1 and 5.2.

8.1. *Reducing the gluing system*

Let us consider equation (5.12) in the gluing system (5.12)-(5.14),

$$\Delta\psi - W_\alpha(x)\psi + (1-\zeta_2)S(\mathtt{w}) + (1-\zeta_1)N(\psi+\zeta_2\phi) + 2\nabla\zeta_1\nabla\phi + \phi\Delta\zeta_1 = 0 \quad \text{in } \mathbb{R}^3 \tag{8.1}$$

where

$$W_\alpha(x) := \left[(1-\zeta_1)f'(u_1) + \zeta_1 H(t) \right].$$

8.1.1. *Solving the linear outer problem*

We consider first the linear problem

$$\Delta\psi - W_\alpha(x)\psi + g(x) \;=\; 0 \quad \text{in } \mathbb{R}^3. \tag{8.2}$$

We observe that globally we have $0 < a < W_\alpha(x) < b$ for certain constants a and b. In fact we can take $a = \min\{\sigma_-^2, \sigma_+^2\} - \tau$ for arbitrarily small $\tau > 0$.

We consider for the purpose the norms for $1 < p \le +\infty$,

$$\|g\|_{p,\mu} := \sup_{x\in\mathbb{R}^3} (1+r(\alpha x))^\mu \|g\|_{L^p(B(x,1))}, \quad r(x',x_3) = |x'|.$$

Lemma 8.1: *Given $p > 3$, $\mu \ge 0$, there is a $C > 0$ such that for all sufficiently small α and any g with $\|g\|_{p,\mu} < +\infty$ there exists a unique ψ solution to Problem (8.2) with $\|\psi\|_{\infty,\mu} < +\infty$. This solution satisfies in addition,*

$$\|D^2\psi\|_{p,\mu} + \|\psi\|_{\infty,\mu} \le C\|g\|_{p,\mu}. \tag{8.3}$$

Proof: We claim that the *a priori* estimate

$$\|\psi\|_{\infty,\mu} \le C\|g\|_{p,\mu} \tag{8.4}$$

holds for solutions ψ with $\|\psi\|_{\infty,\mu} < +\infty$ to problem (8.2) with $\|g\|_{p,\mu} < +\infty$ provided that α is small enough. This and local elliptic estimates in turn implies the validity of (8.3). To see this, let us assume the opposite, namely the existence $\alpha_n \to 0$, and solutions ψ_n to equation (8.2) with $\|\psi_n\|_{\infty,\mu} = 1$, $\|g_n\|_{p,\mu} \to 0$. Let us consider a point x_n with

$$(1 + r(\alpha_n x_n))^\mu \psi_n(x_n) \ge \frac{1}{2}$$

and define

$$\tilde{\psi}_n(x) = (1+r(\alpha_n(x_n+x)))^\mu \psi_n(x_n+x), \quad \tilde{g}_n(x) = (1+r(\alpha_n(x_n+x)))^\mu g_n(x_n+x),$$

$$\tilde{W}_n(x) = W_{\alpha_n}(x_n + x).$$

Then, similarly to what was done in the previous section, we check that the equation satisfied by $\tilde{\psi}_n$ has the form

$$\Delta\tilde{\psi}_n - \tilde{W}_n(x)\tilde{\psi}_n + o(1)\nabla\tilde{\psi}_n + o(1)\tilde{\psi}_n = \tilde{g}_n.$$

$\tilde{\psi}_n$ is uniformly bounded. Then elliptic estimates imply L^∞-bounds for the gradient and the existence of a subsequence uniformly convergent over compact subsets of $\mathbb{R}^3$ to a bounded solution $\tilde{\psi} \neq 0$ to an equation of the form

$$\Delta\tilde{\psi} - W_*(x)\tilde{\psi} = 0 \quad \text{in } \mathbb{R}^3$$

where $0 < a \leq W_*(x) \leq b$. But maximum principle makes this situation impossible, hence estimate (8.4) holds.

Now, for existence, let us consider g with $\|g\|_{p,\mu} < +\infty$ and a collection of approximations g_n to g with $\|g_n\|_{\infty,\mu} < +\infty$, $g_n \to g$ in $L^p_{loc}(\mathbb{R}^3)$ and $\|g_n\|_{p,\mu} \leq C\|g\|_{p,\mu}$. The problem

$$\Delta\psi_n - W_n(x)\psi_n = g_n \quad \text{in } \mathbb{R}^3$$

can be solved since this equation has a positive supersolution of the form $C_n(1 + r(\alpha x))^{-\mu}$, provided that α is sufficiently small, but independently of n. Let us call ψ_n the solution thus found, which satisfies $\|\psi_n\|_{\infty,\mu} < +\infty$. The *a priori* estimate shows that

$$\|D^2\psi_n\|_{p,\mu} + \|\psi_n\|_{\infty,\mu} \leq C\|g\|_{p,\mu}$$

and passing to the local uniform limit up to a subsequence, we get a solution ψ to problem (8.2), with $\|\psi\|_{\infty,\mu} < +\infty$. The proof is complete. $\qquad\square$

8.1.2. *The proof of Lemma 5.1*

Let us call $\psi := \Upsilon(g)$ the solution of Problem (8.2) predicted by Lemma 8.1. Let us write Problem (8.1) as fixed point problem in the space X of $W^{2,p}_{loc}$-functions ψ with $\|\psi\|_X < +\infty$,

$$\psi = \Upsilon(g_1 + K(\psi)) \tag{8.5}$$

where

$$g_1 = (1 - \zeta_2)S(\mathtt{w}) + 2\nabla\zeta_1\nabla\phi + \phi\Delta\zeta_1, \quad K(\psi) = (1 - \zeta_1)N(\psi + \zeta_2\phi).$$

Let us consider a function ϕ defined in $M_\alpha \times \mathbb{R}$ such that $\|\phi\|_{2,p,\mu,\sigma} \leq 1$. Then,

$$| 2\nabla\zeta_1\nabla\phi + \phi\Delta\zeta_1 | \leq Ce^{-\sigma\frac{\delta}{\alpha}} (1 + r(\alpha x))^{-\mu} \|\phi\|_{2,p,\mu,\sigma}.$$

We also have that $\|S(\mathbf{w})\|_{p,\mu,\sigma} \leq C\alpha^3$, hence

$$|(1 - \zeta_2)S(\mathbf{w})| \leq Ce^{-\sigma\frac{\delta}{\alpha}} (1 + r(\alpha x))^{-\mu}$$

and

$$\|g_1\|_{p,\mu} \leq Ce^{-\sigma\frac{\delta}{\alpha}}.$$

Let consider the set

$$\Lambda = \{\psi \in X \ / \ \|\psi\|_X \leq Ae^{-\sigma\frac{\delta}{\alpha}}\},$$

for a large number $A > 0$. Since

$$| K(\psi_1) - K(\psi_2) | \leq C(1 - \zeta_1) \sup_{t\in(0,1)} |t\psi_1 + (1 - t)\psi_2 + \zeta_2\phi| \, |\psi_1 - \psi_2|,$$

we find that

$$\| K(\psi_1) - K(\psi_2) \|_{\infty,\mu} \leq C e^{-\sigma\frac{\delta}{\alpha}} \| \psi_1 - \psi_2 \|_{\infty,\mu}$$

while $\|K(0)\|_{\infty,\mu} \leq C e^{-\sigma\frac{\delta}{\alpha}}$. It follows that the right hand side of equation (8.5) defines a contraction mapping of Λ, and hence a unique solution $\psi = \Psi(\phi) \in \Lambda$ exists, provided that the number A in the definition of Λ is taken sufficiently large and $\|\phi\|_{2,p,\mu,\sigma} \leq 1$. In addition, it is direct to check the Lipschitz dependence of Ψ (5.18) on $\|\phi\|_{2,p,\mu,\sigma} \leq 1$.

Thus, we replace replace $\psi = \Psi(\phi)$ into the equation (5.14) of the gluing system (5.12)-(5.14) and get the (nonlocal) problem,

$$\partial_{tt}\phi + \Delta_{y,M_\alpha}\phi = -\tilde{S}(u_1) - \mathrm{N}(\phi) \quad \text{in } M_\alpha \times \mathbb{R} \tag{8.6}$$

where

$$\mathrm{N}(\phi) := \underbrace{\mathrm{B}(\phi) + [f'(u_1) - f'(w)]\phi}_{\mathrm{N}_1(\phi)} + \underbrace{\zeta_1(f'(u_1) - H(t))\Psi(\phi)}_{\mathrm{N}_2(\phi)} + \underbrace{\zeta_1 N(\Psi(\phi) + \phi)}_{\mathrm{N}_3(\phi)}, \tag{8.7}$$

which is what we concentrate in solving next.

8.2. Proof of Proposition 5.2

We recall from Secton 5.2 that Proposition 5.2 refers to solving the projected problem

$$
\partial_{tt}\phi + \Delta_{y,M_\alpha}\phi = -\tilde{S}(u_1) - \mathbb{N}(\phi) + c(y)w'(t) \quad \text{in } M_\alpha \times \mathbb{R},
$$
$$
\int_{\mathbb{R}} \phi(y,t)\, w'(t)\, dt = 0, \quad \text{for all } \ y \in M_\alpha,
\tag{8.8}
$$

and then adjust h_1 so that $c(y) \equiv 0$. Let $\phi = T(g)$ be the linear operator providing the solution in Proposition 5.1. Then Problem (8.8) can be reformulated as the fixed point problem

$$
\phi = T(-\tilde{S}(u_1) - \mathbb{N}(\phi)) =: \mathcal{T}(\phi), \quad \|\phi\|_{2,p,\mu,\sigma} \le 1
\tag{8.9}
$$

which is equivalent to

$$
\phi = T(-\tilde{S}(u_1) + \alpha^2 \Delta h_1\, w' - \mathbb{N}(\phi)), \quad \|\phi\|_{2,p,\mu,\sigma} \le 1,
\tag{8.10}
$$

since the term added has the form $\rho(y)w'$ which thus adds up to $c(y)w'$. The reason to absorb this term is that because of assumption (5.7), $\|\alpha^2 \Delta h_1\, w'\|_{p,4,\sigma} = O(\alpha^{3-\frac{2}{p}})$ while the remainder has *a priori* size slightly smaller, $O(\alpha^3)$.

8.2.1. Lipschitz character of $\mathbb{N}$

We will solve Problem (8.10) using contraction mapping principle, so that we need to give account of a suitable Lipschitz property for the operator $\mathcal{T}$. We claim the following:

Claim. *We have that for a certain constant $C > 0$ possibly depending on K in (5.7) but independent of $\alpha > 0$, such that for any ϕ_1, ϕ_2 with*

$$
\|\phi_l\|_{2,p,\mu,\sigma} \le K\alpha^3,
$$

$$
\|\mathbb{N}(\phi_1) - \mathbb{N}(\phi_2)\|_{p,\mu+1,\sigma} \le C\alpha\, \|\phi_1 - \phi_2\|_{2,p,\mu,\sigma}
\tag{8.11}
$$

where the operator $\mathbb{N}$ is defined in (8.7).

We study the Lipschitz character of the operator $\mathbb{N}$ through analyzing each of its components. Let us start with N_1. This is a second-order linear operator with coefficients of order α plus a decay of order at least $O(r_\alpha^{-1})$. We recall that $\mathsf{B} = \zeta_2 B$ where in coordinates

$$B = (f'(u_1) - f'(w)) - \alpha^2[(t+h_1)|A|^2 + \Delta_M h_1]\partial_t - 2\alpha\, a_{ij}^0 \partial_j h \partial_{it}$$

$$+ \alpha(t+h)\,[a_{ij}^1 \partial_{ij} - \alpha\, a_{ij}^1(\partial_j h \partial_{it} + \partial_i h \partial_{jt}) + \alpha(b_i^1 \partial_i - \alpha b_i^1 \partial_i h \partial_t))]$$

$$+ \alpha^3 (t+h)^2 b_3^1 \partial_t + \alpha^2 [a_{ij}^0 + \alpha(t+h)a_{ij}^1)]\partial_i h \partial_j h\, \partial_{tt} \qquad (8.12)$$

where, we recall,

$$a_{ij}^1 = O(r_\alpha^{-2}), \quad a_{ij}^1 = O(r_\alpha^{-2}), \quad b_i^1 = O(r_\alpha^{-3}), \quad b_i^3 = O(r_\alpha^{-6}),$$

$$f'(u_1) - f'(w) = O(\alpha^2 r_\alpha^{-2} e^{-\sigma|t|}) \quad \partial_j h = O(r_\alpha^{-1}), \quad |A|^2 = O(r_\alpha^{-4}).$$

We claim that

$$\|\mathsf{N}_1(\phi)\|_{p,\mu+1,\sigma} \le C\,\alpha\,\|\phi\|_{2,p,\mu,\sigma}. \qquad (8.13)$$

The only term of $N_1(\phi)$ that requires a bit more attention is $\alpha^2(\Delta h_1)(\alpha y)\partial_t \phi$. We have

$$\int_{B((y,t),1)} |\alpha^2(\Delta h_1)(\alpha z)\partial_t \phi|^p \, dV_\alpha(z)\, d\tau$$

$$\le C\alpha^{2p}\|\partial_t \phi\|_{L^\infty(B((y,t),1))}(1+r_\alpha(y))^{-4p+4}\int_{B((y,t),1)}|(1+r_\alpha(z))^{4-\frac{4}{p}}(\Delta h_1)(\alpha z)|^p\,dV_\alpha(z)$$

$$\le C\alpha^{2p-2}\|\Delta h_1\|_{L^p(M)}^p e^{-p\sigma|t|}(1+r_\alpha(y))^{-p\mu-4p+4}\|\nabla\phi\|_{\infty,\mu,\sigma},$$

and hence in particular for $p \ge 3$,

$$\|\alpha^2(\Delta h_1)(\alpha y)\partial_t \phi\|_{p,\mu+2,\sigma} \le C\,\alpha^{2-\frac{2}{p}}\|h_1\|_* \|\phi\|_{2,p,\mu,\sigma} \le C\,\alpha^{3-\frac{2}{p}}\|\phi\|_{2,p,\mu,\sigma}.$$

Let us consider now functions ϕ_l with

$$\|\phi_l\|_{2,p,\mu,\sigma} \le 1, \quad l = 1,2.$$

Now, according to Lemma 5.1, we get that

$$\|\mathsf{N}_2(\phi_1) - \mathsf{N}_2(\phi_2)\|_{p,\mu,\sigma} \le C\,e^{-\sigma\frac{\delta}{\alpha}}\|\phi_1 - \phi_2\|_{p,\mu,\sigma}. \qquad (8.14)$$

Finally, we also have that

$$|\mathsf{N}_3(\phi_1) - \mathsf{N}_3(\phi_2)|$$

$$\leq C\zeta_1 \sup_{t\in(0,1)} |t(\Psi(\phi_1)+\phi_1)+(1-t)(\Psi(\phi_2)+\phi_2)|[|\phi_1-\phi_2|+|\Psi(\phi_1)-\Psi(\phi_2)|],$$

hence

$$\|\mathbb{N}_3(\phi_1)-\mathbb{N}_3(\phi_2)\|_{p,2\mu,\sigma} \leq C(\|\phi_1\|_{\infty,\mu,\sigma}+\|\phi_2\|_{\infty,\mu,\sigma}+e^{-\sigma\frac{\delta}{\alpha}})\|\phi_1-\phi_2\|_{\infty,\mu,\sigma}. \tag{8.15}$$

From (8.13), (8.14) and (8.15), inequality (8.11) follows. The proof of the claim is concluded.

8.2.2. *Conclusion of the proof of Proposition 5.2*

The first observation is that choosing $\mu \leq 3$, we get

$$\|\tilde{S}(u_1)+\alpha^2\Delta h_1 w'\|_{p,\mu,\sigma} \leq C\alpha^3. \tag{8.16}$$

Let us assume now that $\phi_1, \phi_2 \in B_\alpha$ where

$$B_\alpha = \{\phi \;/\; \|\phi\|_{2,p,\mu,\sigma} \leq K\alpha^3\}$$

where K is a constant to be chosen. Then we observe that for small α

$$\|\mathbb{N}(\phi)\|_{p,\mu+1,\sigma} \leq C\alpha^4, \quad \text{for all} \quad \phi \in B_\alpha,$$

where C is independent of K. Then, from relations (8.16)-(8.15) we see that if K is fixed large enough independent of α, then the right hand side of equation (8.5) defines an operator that applies B_α into itself, which is also a contraction mapping of B_α endowed with the norm $\| \; \|_{p,\mu\sigma}$, provided that $\mu \leq 3$. We conclude, from contraction mapping principle, the existence of ϕ as required.

The Lipschitz dependence (5.27) is a consequence of series of lengthy but straightforward considerations of the Lipschitz character in h_1 of the operator in the right hand side of equation (8.5) for the norm $\| \; \|_*$ defined in (5.34). Let us recall expression (8.12) for the operator B, and consider as an example, two terms that depend linearly on h_1:

$$A(h_1,\phi) := \alpha\, a_{ij}^0\, \partial_j h_1 \partial_{it}\phi.$$

Then

$$|A(h_1,\phi)| \leq C\alpha|\partial_j h_1|\,|\partial_{it}\phi|.$$

Hence

$$\|A(h_1,\phi)\|_{p,\mu+2,\sigma} \leq C\alpha\|(1+r_\alpha^2)\partial_j h_1\|_\infty\|\partial_{it}\phi\|_{p,\mu,\sigma} \leq C\alpha^4\|h_1\|_*\|\phi\|_{2,p,\mu,\sigma}.$$

Similarly, for $A(\phi, h_1) = \alpha^2 \Delta_M h_1 \, \partial_t \phi$ we have

$$|A(\phi, h_1)| \leq C\alpha^2 |\Delta_M h_1(\alpha y)| (1 + r_\alpha)^{-\mu} e^{-\sigma|t|} \|\phi\|_{2,p,\mu,\sigma}.$$

Hence

$$\|\alpha^2 \Delta_M h_1 \, \partial_t \phi\|_{p,\mu+2,\sigma} \leq C\alpha^{5-\frac{2}{p}} \|h_1\|_* \|\phi\|_{2,p,\mu,\sigma}.$$

We should take into account that some terms involve nonlinear, however mild dependence, in h_1. We recall for instance that $a_{ij}^1 = a_{ij}^1(\alpha y, \alpha(t + h_0 + h_1))$. Examining the rest of the terms involved we find that the whole operator $\mathbb{N}$ produces a dependence on h_1 which is Lipschitz with small constant, and gaining decay in r_α,

$$\|\mathbb{N}(h_1, \phi) - \mathbb{N}(h_2, \phi)\|_{p,\mu+1,\sigma} \leq C\alpha^2 \|h_1 - h_2\|_* \|\phi\|_{2,p,\mu,\sigma}. \tag{8.17}$$

Now, in the error term

$$\mathcal{R} = -\tilde{S}(u_1) + \alpha^2 \Delta h_1 w',$$

we have that

$$\|\mathcal{R}(h_1) - \mathcal{R}(h_2)\|_{p,3,\sigma} \leq C\alpha^2 \|h_1 - h_2\|_*. \tag{8.18}$$

To see this, again we go term by term in expansion (5.15). For instance the linear term $\alpha^2 \, a_{ij}^0 \partial_i h_0 \partial_j h_1 \, w''$. We have

$$|\alpha^2 \, a_{ij}^0 \, \partial_i h_0 \partial_j h_1| \leq C\alpha^2 (1 + r_\alpha)^{-3} e^{-\sigma|t|} \|h_1\|_*$$

so that

$$\|\alpha^2 \, a_{ij}^0 \, \partial_i h_0 \, \partial_j h_1\|_{p,3,\sigma} \leq C\alpha^2 \|h_1\|_*,$$

the remaining terms are checked similarly.

Combining estimates (8.17), (8.18) and the fixed point characterization (8.5) we obtain the desired Lipschitz dependence (5.27) of Φ.

This concludes the proof.

9. The Reduced Problem: Proof of Proposition 5.4

In this section, we prove Proposition 5.4 based on the linear theory provided by Proposition 5.3. Thus, we want to solve the problem

$$\mathcal{J}(h_1) = \Delta_M h_1 + h_1 |A|^2 = G(h_1) + \sum_{i=1}^{J} \frac{c_i}{1 + r^4} \hat{z}_i \quad \text{in } M, \tag{9.1}$$

$$\int_M \frac{h_1 \hat{z}_i}{1 + r^4} \, dV = 0 \quad \text{for all} \quad i = 1, \dots, J,$$

where the linearly independent Jacobi fields $\hat{z}_i$ will be chosen in (10.1) and (10.2) of Section 8, and $G = G_1 + G_2$ was defined in (5.29), (5.30). We will use contraction mapping principle to determine the existence of a unique solution h_1 for which constraint (5.7), namely

$$\|h_1\|_* := \|h_1\|_{L^\infty(M)} + \|(1+r^2)Dh_1\|_{L^\infty(M)} + \|D^2 h_1\|_{p,4-\frac{4}{p}} \leq \mathcal{K}\alpha , \quad (9.2)$$

is satisfied after fixing $\mathcal{K}$ sufficiently large.

We need to analyze the size of the operator G, for which the crucial step is the following estimate.

Lemma 9.1: *Let $\psi(y,t)$ be a function defined in $M_\alpha \times \mathbb{R}$ such that*

$$\|\psi\|_{p,\mu,\sigma} := \sup_{(y,t)\in M_\alpha\times\mathbb{R}} e^{\sigma|t|}(1+r_\alpha^\mu)\,\|\psi\|_{L^p(B((y,t),1))} < +\infty$$

for $\sigma,\mu \geq 0$. The function defined in M as

$$q(y) := \int_{\mathbb{R}} \psi(y/\alpha, t)\, w'(t)\, dt$$

satisfies

$$\|q\|_{p,a} \leq C\,\|\psi\|_{p,\mu,\sigma} \quad (9.3)$$

provided that

$$\mu > \frac{2}{p} + a .$$

In particular, for any $\tau > 0$,

$$\|q\|_{p,2-\frac{2}{p}-\tau} \leq C\,\|\psi\|_{p,2,\sigma} \quad (9.4)$$

and

$$\|q\|_{p,4-\frac{4}{p}} \leq C\,\|\psi\|_{p,4,\sigma} . \quad (9.5)$$

Proof: We have that for $|y| > R_0$

$$\int_{|y|>R_0} |y|^{ap}\left|\int_{\mathbb{R}} \psi(y/\alpha,t)\,w'(t)\,dt\right|^p dV \leq C\int_{\mathbb{R}} w'(t)\,dt \int_{|y|>R_0} |y|^{ap}\,|\psi(y/\alpha,t)|^p\,dV .$$

Now

$$\int_{|y|>R_0} |y|^{ap}\,|\psi(y/\alpha,t)|^p\,dV = \alpha^{ap+2}\int_{|y|>R_0/\alpha} |y|^{ap}\,|\psi(y,t)|^p\,dV_\alpha$$

and

$$\int_{|y|>R_0/\alpha} |y|^{ap}\,|\psi(y,t)|^p\,dV_\alpha \leq C\sum_{i\geq[R_0/\alpha]} i^{ap}\int_{i<|y|<i+1} |\psi(y,t)|^p\,dV_\alpha .$$

Now, $i < |y| < i+1$ is contained in $O(i)$ balls with radius one centered at points of the annulus, hence

$$\int_{i<|y|<i+1} |\psi(y,t)|^p \, dV_\alpha \leq C e^{-\sigma p|t|} i^{1-\mu p} \|\psi\|_{p,\mu}^p$$

$$\leq C e^{-\sigma p|t|} \|\psi\|_{p,\mu}^p \int_{i<|y|<i+1} (1+r_\alpha)^{-\mu p} dV_\alpha$$

$$\leq C e^{-\sigma p|t|} \|\psi\|_{p,\mu}^p \int_{i<|y|<i+1} |\alpha y|^{-\mu p} dV_\alpha$$

$$\leq C e^{-\sigma p|t|} \|\psi\|_{p,\mu}^p \alpha^{-\mu p} i^{1-\mu p}.$$

Then we find

$$\| |y|^a q \|_{L^p(|y|>R_0)}^p \leq C \alpha^{ap-\mu p+2} \|\psi\|_{p,\mu}^p \sum_{i \geq [R_0/\alpha]} i^{ap-\mu p+1}.$$

The sum converges if $\mu > \frac{2}{p} + a$ and in this case

$$\| |y|^a q \|_{L^p(|y|>R_0)}^p \leq C \alpha^{ap-\mu p+2} \alpha^{-ap+\mu p-2} \|\psi\|_{p,\mu}^p = C \|\psi\|_{p,\mu}^p$$

so that

$$\| |y|^a q \|_{L^p(|y|>R_0)} \leq C \|\psi\|_{p,\mu}.$$

Now, for the inner part $|y| < R_0$ in M, the weights play no role. We have

$$\int_{|y|<R_0} |\psi(y/\alpha, t)|^p \, dV = \alpha^2 \int_{|y|<R_0/\alpha} |\psi(y,t)|^p \, dV_\alpha$$

$$\leq C\alpha^2 \sum_{i \leq R_0/\alpha} \int_{i<|y|<i+1} |\psi(y,t)|^p \, dV_\alpha \leq C\alpha^2 \|\psi\|_{p,\mu}^p e^{-\sigma p|t|} \sum_{i \leq R_0/\alpha} i$$

$$\leq C \|\psi\|_{p,\mu}^p e^{-\sigma p|t|}.$$

Hence if $\mu > \frac{2}{p} + a$ we finally get

$$\|q\|_{p,a} \leq C \|\psi\|_{p,\mu}$$

and the proof of (9.3) is concluded. Letting $(\mu, a) = (2, 2 - \frac{2}{p} - \tau)$, $(\mu, a) = (4, 4 - \frac{4}{p})$ respectively in (9.3), we obtain (9.4) and (9.5). $\qquad\square$

Let us apply this result to $\psi(y,t) = \mathtt{N}(\Phi(h_1))$ to estimate the size of the operator G_2 in (5.30). For $\phi = \Phi(h_1)$ we have that

$$G_2(h_1)(y) := c_*^{-1}\alpha^{-2} \int_{\mathbb{R}} \mathtt{N}(\phi)(y/\alpha,t)\, w'\, dt$$

satisfies

$$\|G_2(h_1)\|_{p,4-\frac{4}{p}} \leq C\alpha^{-2}\|\mathtt{N}(\phi)\|_{p,4,\sigma} \leq C\alpha^2.$$

On the other hand, we have that, similarly, for $\phi_l = \Phi(h_l)$, $l = 1,2$,

$$\|G_2(h_1) - G_2(h_2)\|_{p,4-\frac{4}{p}} \leq C\alpha^{-2}\|\mathtt{N}(\phi_1,h_1) - \mathtt{N}(\phi_2,h_2)\|_{p,4,\sigma}.$$

Now,

$$\|\mathtt{N}(\phi_1,h_1) - \mathtt{N}(\phi_1,h_2)\|_{p,4,\sigma} \leq C\alpha^2\|h_1 - h_2\|_*\|\phi_1\|_{2,p,3,\sigma}, \leq C\alpha^5\|h_1 - h_2\|_*,$$

according to inequality (8.17), and

$$\|\mathtt{N}(\phi_1,h_1) - \mathtt{N}(\phi_2,h_1)\|_{p,4,\sigma} \leq C\alpha^2\|\phi_1 - \phi_2\|_{p,3,\sigma} \leq C\alpha^4\|h_1 - h_2\|_*.$$

We conclude then that

$$\|G_2(h_1) - G_2(h_2)\|_{p,4-\frac{4}{p}} \leq C\alpha^2\|h_1 - h_2\|_*.$$

In addition, we also have that

$$\|G_2(0)\|_{p,4-\frac{4}{p}} \leq C\alpha^2$$

for some $C > 0$ possibly dependent of $\mathcal{K}$. On the other hand, it is similarly checked that the remaining small operator $G_1(h_1)$ in (5.29) satisfies

$$\|G_1(h_1) - G_1(h_2)\|_{p,4-\frac{4}{p}} \leq C_1\alpha\|h_1 - h_2\|_*.$$

A simple but crucial observation we make is that

$$c_*G_1(0) = \alpha\, \partial_i h_0 \partial_j h_0 \int_{\mathbb{R}} \zeta_4(t+h_0)a_{ij}^1 w'' w'\, dt + \alpha^{-2} \int_{\mathbb{R}} \zeta_4\, R_1(y,t,0,0)\, w'\, dt$$

so that for a constant C_2 independent of $\mathcal{K}$ in (9.2) we have

$$\|G_1(0)\|_{p,4-\frac{4}{p}} \leq C_2\alpha.$$

In all we have that the operator $G(h_1)$ has an $O(\alpha)$ Lipschitz constant, and in addition satisfies

$$\|G(0)\|_{p,4-\frac{4}{p}} \leq 2C_2\alpha.$$

Let $h = T(g)$ be the linear operator defined by Proposition 5.3. Then we consider the problem (9.1) written as the fixed point problem

$$h_1 = T(G(h_1)), \quad \|h\|_* \le \mathcal{K}\alpha. \tag{9.6}$$

We have

$$\|T(G(h_1))\|_* \le \|T\| \|G(0)\|_{p,4-\frac{4}{p}} + C\alpha\|h_1\|_*.$$

Hence fixing $\mathcal{K} > 2C_2\|T\|$, we find that for all α sufficiently small, the operator TG is a contraction mapping of the ball $\|h\|_* \le \mathcal{K}\alpha$ into itself. We thus have the existence of a unique solution of the fixed problem (9.6), namely a unique solution h_1 to problem (9.1) satisfying (9.2) and the proof of Proposition 5.4 is concluded.

10. Conclusion of the Proof of Theorem 6

We denote in what follows

$$r(x) = \sqrt{x_1^2 + x_2^2}, \quad \hat{r} = \frac{1}{r}(x_1, x_2, 0), \quad \hat{\theta} = \frac{1}{r}(-x_2, x_1, 0).$$

We consider the four Jacobi fields associated to rigid motions, $z_1, \ldots, z_4$ introduced in (2.13). Let J be the number of bounded, linearly independent Jacobi fields of $\mathcal{J}$. By our assumption and the asymptotic expansion of the ends (2.11), $3 \le J \le 4$. (Note that when M is a catenoid, $z_4 = 0$ and $J = 3$.) Let us choose

$$\hat{z}_j = \sum_{l=1}^{4} d_{jl} z_{0l}, j = 1, ..., J \tag{10.1}$$

be normalized such that

$$\int_M q(y)\hat{z}_i\hat{z}_j = 0, \text{ for } i \ne j, \int_M q(y)\hat{z}_i^2 = 1, i, j = 1, \ldots, J. \tag{10.2}$$

In what follows we fix the function q as

$$q(y) := \frac{1}{1 + r(y)^4}. \tag{10.3}$$

So far we have built, for certain constants $\tilde{c}_i$ a solution u of equation (5.36), namely

$$\Delta u + f(u) = \sum_{j=1}^{J} \tilde{c}_i \hat{z}_i(\alpha y) w'(t) q(\alpha y)\zeta_2$$

where u, defined in (5.35) satisfies the following properties

$$u(x) = w(t) + \phi(y, t) \qquad (10.4)$$

near the manifold, meaning this $x = y + (t + h(\alpha y))\nu(\alpha y)$ with

$$y \in M_\alpha, \quad |t| \leq \frac{\delta}{\alpha} + \gamma \log(2 + r(\alpha y)).$$

The function ϕ satisfies in this region the estimate

$$|\phi| + |\nabla\phi| \leq C\alpha^2 \frac{1}{1 + r^2(\alpha y)} e^{-\sigma|t|}. \qquad (10.5)$$

Moreover, we have the validity of the global estimate

$$|\nabla u(x)| \leq \frac{C}{1 + r^3(\alpha x)} e^{-\sigma \frac{\delta}{\alpha}}.$$

We introduce the functions

$$Z_i(x) = \partial_{x_i} u(x), \quad i = 1, 2, 3, \quad Z_4(x) = -\alpha x_2 \partial_{x_2} u + \alpha x_1 \partial_{x_2} u.$$

From the expansion (10.4) we see that

$$\nabla u(x) = w'(t)\nabla t + \nabla\phi.$$

Now, $t = z - h(\alpha y)$ where z designates normal coordinate to M_α. Since $\nabla z = \nu = \nu(\alpha y)$ we then get

$$\nabla t = \nu(\alpha y) - \alpha \nabla h(\alpha y).$$

Let us recall that h satisfies $h = (-1)^k \beta_k \log r + O(1)$ along the k-th end, and

$$\nabla h = (-1)^k \frac{\beta_k}{r} \hat{r} + O(r^{-2}).$$

From estimate (10.5) we find that

$$\nabla u(x) = w'(t)(\nu - \alpha(-1)^k \frac{\beta_k}{r_\alpha} \hat{r}) + O(\alpha r_\alpha^{-2} e^{-\sigma|t|}). \qquad (10.6)$$

From here we get that near the manifold,

$$Z_i(x) = w'(t)(z_i(\alpha y) - \alpha(-1)^k \frac{\beta_k}{r_\alpha} \hat{r} e_i) + O(\alpha r_\alpha^{-2} e^{-\sigma|t|}), \quad i = 1, 2, 3, \quad (10.7)$$

$$Z_4(x) = w'(t) z_{04}(\alpha y) + O(\alpha r_\alpha^{-1} e^{-\sigma|t|}). \qquad (10.8)$$

Using the characterization (5.36) of the solution u and barriers (in exactly the same way as in Lemma 10.1 of [20] which estimates eigenfunctions of the linearized operator), we find the following estimate for $r_\alpha(x) > R_0$:

$$|\nabla u(x)| \le C \sum_{k=1}^{m} e^{-\sigma|x_3 - \alpha^{-1}(F_k(\alpha x') + \beta_j \alpha \log |\alpha x'|)|} . \tag{10.9}$$

We claim that

$$\int_{\mathbb{R}^3} (\Delta u + f(u)) Z_i(x) \, dx = 0 \quad \text{for all} \quad i = 1, \dots, 4 \tag{10.10}$$

so that

$$\sum_{j=1}^{J} \tilde{c}_j \int_{\mathbb{R}^3} q(\alpha x) \hat{z}_j(\alpha y) w'(t) Z_i(x) \zeta_2 \, dx = 0 \quad \text{for all} \quad i = 1, \dots, 4. \tag{10.11}$$

Let us accept this fact for the moment. Let us observe that from estimates (10.7) and (10.8),

$$\alpha^2 \int_{\mathbb{R}^3} q(\alpha x) \hat{z}_j(\alpha y) w'(t) \sum_{l=1}^{4} d_{il} Z_l(x) \zeta_2 \, dx = \int_{-\infty}^{\infty} w'(t)^2 dt \int_{M} q \, \hat{z}_j \hat{z}_i dV + o(1)$$

with $o(1)$ is small with α. Since the functions $\hat{z}_i$ are linearly independent on any open set because they solve an homogeneous elliptic PDE, we conclude that the matrix with the above coefficients is invertible. Hence from (10.11) and (10.2), all $\tilde{c}_i$'s are necessarily zero. We have thus found a solution to the Allen Cahn equation (2.1) with the properties required in Theorem 6.

It remains to prove identities (10.10). The idea is to use the invariance of $\Delta + f(u)$ under rigid translations and rotations. This type of Pohozaev identity argument has been used in a number of places, see for instance [31].

In order to prove that the identity (10.10) holds for $i = 3$, we consider a large number $R \gg \frac{1}{\alpha}$ and the infinite cylinder

$$C_R = \{x \ / \ x_1^2 + x_2^2 < R^2\}.$$

Since in C_R the quantities involved in the integration approach zero at exponential rate as $|x_3| \to +\infty$ uniformly in (x_1, x_2), we have that

$$\int_{C_R} (\Delta u + f(u)) \partial_{x_3} u - \int_{\partial C_R} \nabla u \cdot \hat{r} \, \partial_{x_3} u = \int_{C_R} \partial_{x_3} \left(F(u) - \frac{1}{2} |\nabla u|^2 \right) = 0.$$

We claim that

$$\lim_{R \to +\infty} \int_{\partial C_R} \nabla u \cdot \hat{r} \, \partial_{x_3} u = 0.$$

Using estimate (10.6) we have that near the manifold,

$$\partial_{x_3} u \nabla u(x) \cdot \hat{r} = w'(t)^2 ((\nu - \alpha(-1)^k \frac{\beta_k}{r_\alpha} \hat{r}) \cdot \hat{r}) \nu_3 + O(\alpha e^{-\sigma|t|} \frac{1}{r^2}).$$

Let us consider the k-th end, which for large r is expanded as

$$x_3 = F_{k,\alpha}(x_1, x_2) = \alpha^{-1}(a_k \log \alpha r + b_k + O(r^{-1}))$$

so that

$$(-1)^k \nu = \frac{1}{\sqrt{1 + |\nabla F_{k,\alpha}|^2}} (\nabla F_{k,\alpha}, -1) = \frac{a_k}{\alpha} \frac{\hat{r}}{r} - e_3 + O(r^{-2}). \quad (10.12)$$

Then on the portion of C_R near this end we have that

$$(\nu - \alpha(-1)^k \frac{\beta_k}{r_\alpha} \hat{r}) \cdot \hat{r} \nu_3 = -\alpha^{-1} \frac{a_k + \alpha\beta_k}{R} + O(R^{-2}). \quad (10.13)$$

In addition, also, for $x_1^2 + x_2^2 = R^2$ we have the expansion

$$t = (x_3 - F_{k,\alpha}(x_1, x_2) - \beta_k \log \alpha r + O(1))(1 + O(R^{-2}))$$

with the same order valid after differentiation in x_3, uniformly in such (x_1, x_2). Let us choose $\rho = \gamma \log R$ for a large, fixed γ. Observe that on ∂C_R the distance between ends is greater than 2ρ whenever α is sufficiently small. We get,

$$\int_{F_{k,\alpha}(x_1,x_2)+\beta_k \log \alpha r - \rho}^{F_{k,\alpha}(x_1,x_2)+\beta_k \log \alpha r + \rho} w'(t)^2 dx_3 = \int_{-\infty}^{\infty} w'(t)^2 dt + O(R^{-2}).$$

Because of estimate (10.9) we conclude, fixing appropriately γ, that

$$\int_{\cap_k \{|x_3 - F_{k,\alpha}| > \rho\}} \partial_{x_3} u \nabla u(x) \cdot \hat{r} \, dx_3 = O(R^{-2}).$$

As a conclusion

$$\int_{-\infty}^{\infty} \partial_{x_3} u \nabla u \cdot \hat{r} \, dx_3 = -\frac{1}{\alpha R} \sum_{k=1}^{m} (a_k + \alpha\beta_k) \int_{-\infty}^{\infty} w'(t)^2 \, dt + O(R^{-2})$$

and hence

$$\int_{\partial C_R} \partial_{x_3} u \nabla u(x) \cdot \hat{r} = -\frac{2\pi}{\alpha} \sum_{k=1}^{m} (a_k + \alpha\beta_k) + O(R^{-1}).$$

But $\sum_{k=1}^{m} a_k = \sum_{k=1}^{m} \beta_k = 0$ and hence (10.10) for $i = 3$ follows after letting $R \to \infty$.

Let us prove the identity for $i = 2$. We need to carry out now the integration against $\partial_{x_2} u$. In this case we get

$$\int_{C_R} (\Delta u + f(u)) \partial_{x_2} u = \int_{\partial C_R} \nabla u \cdot \hat{r} \, \partial_{x_2} u + \int_{C_R} \partial_{x_2} \left(F(u) - \frac{1}{2} |\nabla u|^2 \right).$$

We have that

$$\int_{C_R} \partial_{x_2} \left(F(u) - \frac{1}{2} |\nabla u|^2 \right) = \int_{\partial C_R} \left(F(u) - \frac{1}{2} |\nabla u|^2 \right) n_2$$

where $n_2 = x_2/r$. Now, near the ends estimate (10.6) yields

$$|\nabla u|^2 = |w'(t)|^2 + O(e^{-\sigma |t|} \frac{1}{r^2})$$

and arguing as before, we get

$$\int_{-\infty}^{\infty} |\nabla u|^2 dx_3 = m \int_{-\infty}^{\infty} |w'(t)|^2 dt + O(R^{-2}).$$

Hence

$$\int_{\partial C_R} |\nabla u|^2 n_2 = m \int_{-\infty}^{\infty} |w'(t)|^2 dt \int_{[r=R]} n_2 + O(R^{-1}).$$

Since $\int_{[r=R]} n_2 = 0$ we conclude that

$$\lim_{R \to +\infty} \int_{\partial C_R} |\nabla u|^2 \, n_2 = 0.$$

In a similar way we get

$$\lim_{R \to +\infty} \int_{\partial C_R} F(u) \, n_2 = 0.$$

Since near the ends we have

$$\partial_{x_2} u = w'(t)(\nu_2 - \alpha(-1)^k \frac{\beta_k}{r_\alpha} \hat{r} e_2) + O(\alpha r^{-2} e^{-\sigma |t|})$$

and from (10.12) $\nu_2 = O(R^{-1})$, completing the computation as previously done yields

$$\int_{\partial C_R} \nabla u \cdot \hat{r} \, \partial_{x_2} u = O(R^{-1}).$$

As a conclusion of the previous estimates, letting $R \to +\infty$ we finally find the validity of (10.10) for $i = 2$. Of course the same argument holds for $i = 1$.

Finally, for $i = 4$ it is convenient to compute the integral over C_R using cylindrical coordinates. Let us write $u = u(r, \theta, z)$. Then

$$\int_{C_R} (\Delta u + f(u)) (x_2 \partial_{x_1} u - x_1 \partial_{x_1} u)$$

$$= \int_0^{2\pi} \int_0^R \int_{-\infty}^\infty [u_{zz} + r^{-1}(r u_r)_r + f(u)] u_\theta \, r \, d\theta \, dr \, dz$$

$$= -\frac{1}{2} \int_0^{2\pi} \int_0^R \int_{-\infty}^\infty \partial_\theta [u_z^2 + u_r^2 - 2F(u)] r \, d\theta \, dr \, dz + R \int_{-\infty}^\infty \int_0^{2\pi} u_r \, u_\theta(R, \theta, z) \, d\theta \, dz$$

$$= 0 + \int_{\partial C_R} u_r u_\theta \ .$$

On the other hand, on the portion of ∂C_R near the ends we have

$$u_r \, u_\theta = w'(t)^2 R(\nu \cdot \hat{r})(\nu \cdot \hat{\theta}) + O(R^{-2} e^{-\sigma |t|}).$$

From (10.12) we find

$$(\nu \cdot \hat{r})(\nu \cdot \hat{\theta}) = O(R^{-3}),$$

hence

$$u_r \, u_\theta = w'(t)^2 O(R^{-2}) + O(R^{-2} e^{-\sigma |t|})$$

and finally

$$\int_{\partial C_R} u_r \, u_\theta = O(R^{-1}).$$

Letting $R \to +\infty$ we obtain relation (10.10) for $i = 4$. The proof is concluded.

Acknowledgments

We thank Professors M. Kowalczyk and F. Pacard for useful discussions.

References

1. A. Ambrosetti and A. Malchiodi, *Perturbation methods and semilinear elliptic problems on* $\mathbb{R}^n$. Progress in Mathematics Series, vol. 240 (Birkhäuser, 2006).
2. W. Ao, J. Wei and J. Zeng, "An optimal bound on the number of interior spike solutions for the Lin-Ni-Takagi problem", *J. Funct. Anal.*, **265** (2013), no. 7, 1324–1356.

3. A. Bahri, *Critical points at infinity in some variational problems*. Pitman Research notes in Mathematics 182, 1982.

4. A. Bahri and J.M. Coron, "On a nonlinear elliptic equation involving the critical Sobolev exponent: the effect of the topology of the domain", *Comm. Pure Appl. Math.* **41** (1988), 253–294.

5. A. Bahri, Y.Y. Li and O. Rey, "On a variational problem with lack of compactness: the topological effect of the critical points at infity", *Calc. Var. Partial Differential Equations* **3** (1995), 67–93.

6. Jaeyoung Byeon and Kazunaga Tanaka, "Semiclassical standing waves with clustering peaks for nonlinear Schroedinger equations", *Memoirs of the American Mathematical Society*, to appear.

7. Jaeyoung Byeon and Kazunaga Tanaka, "Semi-classical standing waves for nonlinear Schrödinger equations at structurally stable critical points of the potential", *Journal of the European Mathematical Society*, to appear.

8. E. Bombieri, E. De Giorgi, E. Giusti, "Minimal cones and the Bernstein problem", *Invent. Math.* **7** (1969), 243–268.

9. L. Ambrosio and X. Cabré, "Entire solutions of semilinear elliptic equations in $\mathbb{R}^3$ and a conjecture of De Giorgi", *Journal Amer. Math. Soc.* **13** (2000), 725–739.

10. L. Caffarelli, A. Córdoba, "Uniform convergence of a singular perturbation problem", *Comm. Pure Appl. Math.* **XLVII** (1995), 1–12.

11. L. Caffarelli, A. Córdoba, "Phase transitions: uniform regularity of the intermediate layers", *J. Reine Angew. Math.* **593** (2006), 209–235.

12. C.J. Costa, *Imersoes minimas en $\mathbb{R}^3$ de genero un e curvatura total finita. PhD thesis*, IMPA, Rio de Janeiro, Brasil (1982).

13. C.J. Costa, "Example of a complete minimal immersions in $\mathbb{R}^3$ of genus one and three embedded ends", *Bol. Soc. Bras. Mat.* **15** (1-2)(1984), 47–54.

14. X. Cabré, J. Terra "Saddle-shaped solutions of bistable diffusion equations in all of $\mathbb{R}^{2m}$", *Preprint* 2008.

15. M. del Pino, P. Felmer, "Local mountain passes for semilinear elliptic problems in unbounded domains", *Calc. Var. Partial Differential Equations* **4** (1996), 121–137.

16. M. del Pino, P. Felmer, "Multi-peak bound states of nonlinear Schrdinger equations", *Ann. Inst. H. Poincaré, Analyse Nonlineaire* **15** (1998), 127–149.

17. M. del Pino, P. Felmer, M. Musso, "Two-bubble solutions in the super-critical Bahri-Coron's problem", *Cal. Var. Partial Differential Equations* **16** (2003), 113–145.

18. M. del Pino, M. Kowalczyk and M. Musso, "Singular limits in Liouville-type equations", *Cal. Var. PDE.* **24** (2005), no. 1, 47–81.

19. M. del Pino, M. Kowalczyk and J. Wei, "Concentration on curves for nonlinear Schrödinger equations", *Communications on Pure and Applied Mathematics* **60** (2007), no. 1, 113–146.

20. M. del Pino, M. Kowalczyk and J. Wei, "Entire aolutions of the Allen-Cahn equation and complete embedded minimal surfaces of finite total curvature", *Journal of Differential Geometry* **83** (2013), no. 1, 67–131.

21. M. del Pino, M. Kowalczyk, F. Pacard and J. Wei, "The Toda system and

multiple-end solutions of autonomous planar elliptic problems, *Advances in Mathematics* **224** (2010), no. 4, 1462–1516.

22. M. del Pino, M. Kowalczyk, J. Wei and J. Yang, "Interface foliation near minimal submanifolds in Riemannian manifolds with positive Ricci curvature", *Geom. Funct. Anal.* **20** (2010), no. 4, 918–957.

23. H. Dang, P.C. Fife, L.A. Peletier, "Saddle solutions of the bistable diffusion equation," *Z. Angew. Math. Phys.* **43** (1992), no. 6, 984–998.

24. E. De Giorgi, *Convergence problems for functionals and operators,* Proc. Int. Meeting on Recent Methods in Nonlinear Analysis (Rome, 1978), 131–188, Pitagora, Bologna (1979).

25. M. del Pino, M. Kowalczyk, J. Wei, "On De Giorgi's Conjecture in Dimensions $N \geq 9$", *Annals of Mathematics* **174** (2011), no. 3, 1485–1569.

26. A. Farina and E. Valdinoci, *The state of art for a conjecture of De Giorgi and related questions,* to appear in "Reaction-Diffusion Systems and Viscosity Solutions", World Scientific, 2008.

27. A. Floer and A. Weinstein, "Nonspreading wave packets for the cubic Schrödinger equation with a bounded potential", *J. Funct. Anal.* **69** (1986), 397–408.

28. C. Gui and J. Wei, "Multiple interior peak solutions for some singularly perturbed Neumann problems", *J. Diff. Eqns.* **158** (1999), no. 1, 1–27.

29. N. Ghoussoub and C. Gui, "On a conjecture of De Giorgi and some related problems", *Math. Ann.* **311** (1998), 481–491.

30. D. Gilbarg and N. S. Trudinger, *Elliptic Partial Differential Equations of Second order*, 2nd edition, Springer-Verlag, 1983.

31. C. Gui, "Hamiltonian identities for elliptic partial differential equations", *J. Funct. Anal.* **254** (2008), no. 4, 904–933.

32. R. Gulliver, "Index and total curvature of complete minimal surfaces", *Proc. Symp. Pure Math.* **44** (1986), 207–212.

33. L. Hauswirth and F. Pacard, "Higher genus Riemann minimal surfaces", *Invent. Math.* **169** (2007), 569–620.

34. D. Hoffman and W.H. Meeks III, "A complete embedded minimal surface in $\mathbb{R}^3$ with genus one and three ends", *J. Diff. Geom.* 21 (1985), 109–127.

35. D. Hoffman and W.H. Meeks III, "The asymptotic behavior of properly embedded minimal surfaces of finite topology", *J. Am. Math. Soc.* 4(2) (1989), 667–681.

36. D. Hoffman and W. H. Meeks III. "The strong halfspace theorem for minimal surfaces", *Inventiones Math.* 101 (1990), 373–377.

37. D. Hoffman and W.H. Meeks III, "Embedded minimal surfaces of finite topology", *Ann. Math.* **131** (1990), 1–34.

38. D. Hoffman and H. Karcher, *Complete embedded minimal surfaces of finite total curvature.* In Geometry V, Encyclopaedia Math. Sci., vol. 90, pp. 5–93, 262–272. Springer Berlin (1997).

39. X. Kang, J. Wei, "On interacting bumps of semi-classical states of nonlinear Schrödinger equations", *Adv. Diff. Eqn.* **5** (2000), 899–928.

40. Y.Y. Li, J. Wei and H. Xu, "Multiple-bump solutions of $-\Delta u = K(x)u^{\frac{N+2}{N-2}}$

on lattices in $\mathbb{R}^{n}$", *Crelle Journal*, to appear.

41. Y.Y. Li and L. Nirenberg, "The Dirichlet problem for singularly perturbed elliptic equations", *Comm. Partial Differential Equations* **23** (1998), 487–545.

42. F.H. Lin, W.M. Ni and J.C. Wei, "On the number of solutions for some singularly perturbed Neumann problems", *Comm. Pure Appl. Math.* 60 (2007), no. 1, 113–146.

43. D. Jerison and R. Monneau, "Towards a counter-example to a conjecture of De Giorgi in high dimensions", *Ann. Mat. Pura Appl.* **183** (2004), 439–467.

44. N. Kapuleas, "Complete embedded minimal surfaces of finite total curvature", *J. Differential Geom.* **45** (1997), 95–169.

45. F. J. López, A. Ros, "On embedded complete minimal surfaces of genus zero", *J. Differential Geom.* **33** (1991), no. 1, 293–300.

46. L. Modica, *Convergence to minimal surfaces problem and global solutions of* $\Delta u = 2(u^3 - u)$. Proceedings of the International Meeting on Recent Methods in Nonlinear Analysis (Rome, 1978), pp. 223–244, Pitagora, Bologna, (1979).

47. F. Morabito, *Index and nullity of the Gauss map of the Costa-Hoffman-Meeks surfaces*, preprint, 2008.

48. A. Malchiodi, "Some new entire solutions of semilinear elliptic equations in R^N", *Adv. in Math.* **221** (2009), no. 6, 1843–1909.

49. S. Nayatani, "Morse index and Gauss maps of complete minimal surfaces in Euclidean 3–space", *Comm. Math. Helv.* 68(4) (1993), 511–537.

50. S. Nayatani, "Morse index of complete minimal surfaces". In: Rassis, T.M. (ed.) The Problem of Plateau, pp. 181–189 (1992).

51. R. Osserman, *A survey of minimal surfaces*, Math. Studies 25, Van Nostrand, New York, 1969.

52. F. Pacard and M. Ritoré, "From the constant mean curvature hypersurfaces to the gradient theory of phase transitions," *J. Differential Geom.* **64** (2003), no. 3, 359–423.

53. J. Pérez, A. Ros, "The space of properly embedded minimal surfaces with finite total curvature", *Indiana Univ. Math. J.* **45** (1996), no. 1, 177–204.

54. Oh, Y.-G. "On positive multibump states of nonlinear Schroedinger equation under multiple well potentials," *Commun. Math. Phys.* **131** (1990), 223–253.

55. F. Pacard and M. Ritore, "From the constant mean curvature hypersurfaces to the gradient theory of phase transitions," *J. Differential Geom.* **64** (2003), no. 3, 359–423.

56. O. Rey, J. Wei, "Blowing up solutions for an elliptic Neumann problem with sub- or supercritical nonlinearity. I. N = 3", *J. Funct. Anal.* **212** (2004), 472–499.

57. O. Rey, J. Wei, "Blowing up solutions for an elliptic Neumann problem with sub- or supercritical nonlinearity. II. $N \geq 4$", *Ann. Inst. H. Poincar Anal. Non Linaire* **22** (2005), 459–484.

58. O. Savin, "Regularity of flat level sets in phase transitions", *Annals of Mathematics* **169** (2009), 41–78.

59. R. Schoen, "Uniqueness, symmetry, and embeddedness of minimal surfaces", *J. Differential Geom.* **18** (1983), 791–809.

60. J.E. Hutchinson, Y. Tonegawa, "Convergence of phase interfaces in the van

der Waals-Cahn-Hilliard theory", *Cal. Var. Partial Differential Equations* **10** (2000), no. 1, 49–84.

61. M. Traizet, "An embedded minimal surface with no symmetries", *J. Differential Geom.* **60** (2002) 103–153.

62. J. Wei and M. Winter, *Mathematical Aspects of Pattern Formation in Biological Systems*, Applied Mathematical Sciences Series, Vol. 189, Springer, 2014, ISBN: 978-4471-5525-6.

63. J. Wei and S. Yan, "Infinitely many positive solutions for the nonlinear Schrodinger equations in $\mathbb{R}^N$", *Cal. Var. PDE* **37** (2000), 423–439.

64. J. Wei and S. Yan, "Infinitely many solution for prescribed curvature problem on S^N", *J. Func. Anal.* **258** (2010), 3048–3081.

EINSTEIN CONSTRAINT EQUATIONS ON RIEMANNIAN MANIFOLDS

Quốc Anh Ngô

Department of Mathematics
College of Science, Viêt Nam National University
Hà Nôi, Viêt Nam
bookworm_vn@yahoo.com
nqanh@vnu.edu.vn

Starting from the Einstein equation in general relativity, we carefully derive the Einstein constraint equations which specify initial data for the Cauchy problem for the Einstein equation. Then, we show how to use the conformal method to study these constraint equations.

0. Introduction

It is well-known that the Einstein theory of relativity (or commonly known as general relativity) is a geometric theory of gravitation. In this theory, gravity is considered as a geometric property of space and time. Because of the geometric property of space and time, general relativity partially includes both special relativity and the Newton law of universal gravitation as special cases.

Theoretically, general relativity describes objects in large scale, like the universe, as Lorentzian manifolds on which gravitation interacts and the universe evolves over time through a system of partial differential equations known as the Einstein equations. Being the central object of the theory, studying the Einstein equations becomes a significant subject in order to understand the whole theory.

In an effort to solve the general Einstein equations, physicists first try to tackle the equations in some simple cases. Fortunately, some remarkably solutions have been found in this direction. Although general relativity nearly coincides with the Newton law of universal gravitation, those known solutions for the Einstein equations in particular cases have led theoretical physicists to predict some new phenomena which deserve investigation carefully. However, much less is known about the solutions of the general Einstein equations. On the other hand, due to the geometric nature of the theory, solving the general Einstein equations turns out to be a wonderful research topic not only for physicists but also for mathematicians, pushing the development of the research rapidly.

However, along with the rapid development of the research, it poses many challenging problems to mathematicians, for example, the initial value problems, the well-posedness problems, the global stability problems, etc. Among these problems, the initial value problems which have their own interest from the mathematical point of view turns out to be the most interesting problem since it involves the theoretical question of the beginning and the end of our universe. When solving the initial value problems, one needs to solve the so-called constraint equations that form an underdetermined system of equations which are not easy to solve. For this reason, understanding the constraint equations is a key step to understanding the initial value problems.

The structure of this lecture notes is as follows:

Contents

We would like to mention that the first four sections are an exposition of the well-known foundation of basics of general relativity and the Einstein equation. Among lots of sources in the literature that we can borrow from, we mainly follow an interesting lecture notes by Corvino in [Cor] and a pedagogical approach by Gourgoulhon in [Gou12]. In addition, we also benefit knowledge from [CBru09] by Choquet-Bruhat and from [Tay11] by Taylor. Other excellent works related to general relativity and the Einstein equation can be found, for example, in [BI04, CGP10].

1. Geometry of Spacetimes in General Relativity

1.1. *Basics of differential geometry*

One of the main subjects in studies of differential geometry in general and of Riemannian geometry in particular are Riemannian manifolds. Loosely speaking, $\mathcal{M}$ is called a (smooth) Riemannian manifold of dimension $n+1$ if:

- $\mathcal{M}$ is a real (smooth) manifold of dimension $n+1$ equipped with
- an inner product $\overline{g}$ on the tangent space $T_P\mathcal{M}$ at any point $P \in \mathcal{M}$ which is positive definite in the sense $\overline{g}(X,X) \geqslant C|X|^2$ for some positive constant C and
- the map $P \mapsto \overline{g}(X(P), Y(P))$ is smooth.

A special case of great importance to general relativity is a Lorentzian manifold, a special case of a wider class of manifolds, called pseudo-Riemannian manifolds, known as a generalization of Riemannian manifolds. Roughly speaking, $\mathcal{M}$ is called a pseudo-Riemannian manifold if

- it is a generalization of a Riemannian manifold in which the metric tensor $\overline{g}$ need not be positive-definite;
- instead a weaker condition of non-degeneracy is imposed on $\overline{g}$: $\nexists X \in T_P\mathcal{M}$ with $X \neq 0$ such that $\overline{g}(X, Y) = 0$ for all $Y \in T_P\mathcal{M}$.

By the Sylvester law of inertia applied to the quadratic form $\overline{g}(x, x)$, we know that the signature (p, q, r) of the metric tensor $\overline{g}$ is invariant with $p + q + r = n + 1$ where by the pair (p, q, r) we mean positive, negative, and zeros. In this context, by pseudo-Riemannian metrics we mean those $\overline{g}$ having the signature $(p, q, 0)$ where p, q are non-negative. Particularly, in the case $p = n$ and therefore $q = 1$, such a metric $\overline{g}$ is called a Lorentzian metric and $(\mathcal{M}, \overline{g})$ is called a Lorentzian manifold. Throughout this notes, we repeatedly use $g(\cdot, \cdot)$ and $\langle \cdot, \cdot \rangle_g$ to denote the metric. Sometimes, we use $\langle X, Y \rangle_{\overline{g}}$ or $X \cdot Y$ instead of $\overline{g}(X, Y)$ if no confusion occurs.

As we shall see right now, lots of geometric quantities can be defined on pseudo-Riemannian manifolds. Given a pseudo-Riemannian manifold $(\mathcal{M}, \overline{g})$, there is an affine connection $\overline{\nabla}$. Then we can define the so-called Riemann curvature tensor $\overline{\mathrm{Rm}}$ in terms of the connection $\overline{\nabla}$ as follows

$$\overline{\mathrm{Rm}}(X, Y, Z) = \overline{\mathrm{Rm}}(X, Y)Z = \overline{\nabla}_X\overline{\nabla}_Y Z - \overline{\nabla}_Y\overline{\nabla}_X Z - \overline{\nabla}_{[X,Y]}Z. \quad (1.1)$$

Using index notation, we can write

$$\overline{\mathrm{Rm}} = \overline{\mathrm{Rm}}^l_{ijk} \frac{\partial}{\partial x^l} \otimes dx^i \otimes dx^j \otimes dx^k$$

and by lowering the index l we also obtain

$$\overline{\mathrm{Rm}}_{ijkl} = \overline{\mathrm{Rm}}^m_{ijk}\overline{g}_{ml} = \left\langle \overline{\mathrm{Rm}}\left(\frac{\partial}{\partial x^i}, \frac{\partial}{\partial x^j}, \frac{\partial}{\partial x^k}\right), \frac{\partial}{\partial x^l} \right\rangle_{\overline{g}}.$$

The Ricci tensor, denoted by Ric, is then defined to be the trace of the Riemann curvature tensor Rm with respect to the first and the last indexes, i.e.

$$\overline{\mathrm{Ric}}(Y, Z) = \mathrm{tr}\left(X \mapsto \mathrm{Rm}(X, Y)Z\right).$$

Therefore, in local coordinates, we write

$$\overline{\mathrm{Ric}} = \mathrm{Ric}_{\overline{g}} = \overline{\mathrm{Ric}}_{jk} = \overline{\mathrm{Rm}}^i_{ijk}. \quad (1.2)$$

Taking the trace of the Ricci tensor Ric, we obtain the scalar curvature, denoted by Scal, that is to say

$$\overline{\mathrm{Scal}} = \mathrm{Scal}_{\overline{g}} = \overline{g}^{jk}\overline{\mathrm{Ric}}_{jk}, \quad (1.3)$$

in local coordinates.

1.2. *What are spacetimes?*

In general relativity, the main objects of study are spacetimes; more precisely, a spacetime in general relativity is nothing but a Lorentzian manifold. An usual spacetime manifold is of dimension four, but higher dimensions are considered in the aim of unification of gravitation with the other fundamental forces of nature, electromagnetism, weak and strong interactions, and also in super-symmetric theories. Throughout this notes, we always assume that spacetimes are of dimension $n + 1$.

Let $(\mathcal{M}, \bar{g})$ be a spacetime, recall from Subsection 1.1 that $(\mathcal{M}, \bar{g})$ is a differentiable manifold of dimension $n + 1$ equipped with a non-degenerate, smooth, symmetric metric tensor $\bar{g}$. A hypersurface of $\mathcal{M}$ is the image Σ of some n-dimensional manifold M by an embedding (meaning a homeomorphism or one-to-one mapping such that both directions are continuous)

$$\Phi : M \to \mathcal{M}.$$

Clearly, the one-to-one property guarantees that Σ does not "intersect itself".

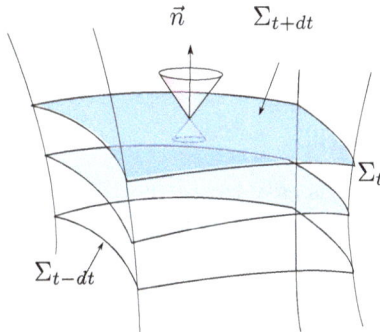

Fig. 1.   The spacetime foliation.

Observe that any hyper-surface can be defined locally as the set of points for which a scalar field on $\mathcal{M}$ is constant. Denoting this scalar field by t and setting the constant to 0, we obtain from the above discussion

$$\forall P \in \mathcal{M}, \quad P \in \Sigma \quad \text{if and only if} \quad t(P) = 0.$$

In other words, the hypersurface Σ is the zero level set of t.

Next, we denote by γ the induced metric of $\bar{g}$ onto the hypersurface Σ. Then Σ is said to be:

- *spacelike*: if γ is Riemannian, i.e. it has signature $(+,+,\dots,+)$.
- *timelike*: if γ is Lorentzian, i.e. it has signature $(-,+,\dots,+)$.
- *null*: if γ is degenerate, i.e. it has signature $(0,+,\dots,+)$.

Since Σ is a level set of t, the gradient 1-form $\overline{\nabla} t = (\overline{\nabla}_\beta t)dx^\beta$ is normal to Σ in the following sense

$$\forall v \in T\Sigma, \quad \overline{\nabla} t(v) = 0.$$

Hence its dual $\overline{\nabla}^\alpha t = \bar{g}^{\alpha\beta}\overline{\nabla}_\beta t$ is a vector field normal to Σ and satisfies:

- $\overline{\nabla} t$ is timelike if and only if Σ is spacelike.
- $\overline{\nabla} t$ is spacelike if and only if Σ is timelike.
- $\overline{\nabla} t$ is null if and only if Σ is null.

In case Σ is not null, we can normalize $\overline{\nabla} t$ to make it unit, then denote by $\vec{n}$:

$$\vec{n} = \frac{1}{\sqrt{\pm\overline{\nabla} t \cdot \overline{\nabla} t}}\overline{\nabla} t \qquad (1.4)$$

with the sign $+$ for a timelike Σ and the sign $-$ for a spacelike one. The coefficient $1/(\pm\overline{\nabla} t \cdot \overline{\nabla} t)^{1/2}$ is called the lapse function, usually denoted by N. Hence, the normal vector field $\vec{n}$ satisfies

- $\vec{n} \cdot \vec{n} = -1$ if Σ is spacelike.
- $\vec{n} \cdot \vec{n} = 1$ if Σ is timelike.

For any X tangent to M, it is easy to see that $\langle \overline{\nabla}_X \vec{n}, \vec{n} \rangle = 0$; hence $\overline{\nabla}_X \vec{n}$ is tangent to Σ. Similarly, vector X is called

- *spacelike*: if $X \cdot X > 0$.
- *timelike*: if $X \cdot X < 0$.
- *null*: if $X \cdot X = 0$.

As being used, for spacetimes (of dimension $n+1$), we use bar " $\overline{\cdot}$ " to denote geometric quantities, like $\overline{\mathrm{Rm}}$, $\mathrm{Scal}_{\bar{g}}$, $\overline{\Gamma}^k_{ij}$, etc, while for geometric quantities for spacelike, we simply drop bars, for example, Rm, Scal_g, Γ^k_{ij}, etc.

To setup a suitable spacetime configuration, we need further conditions. Σ is a Cauchy surface if Σ is a spacelike hypersurface such that each causal (for example, timelike or null) curve without end point only intersects Σ at once and only once. (Equivalently, Σ is a Cauchy surface if and only if its domain of dependence is the whole spacetime $\mathcal{M}$.)

Be careful, not all spacetimes admit a Cauchy surface, for those that do, we call them globally hyperbolic spacetimes. As mentioned in [CBru09], the word "globally hyperbolic" comes from the fact that the wave equation is well-posed in these spacetimes, see also [Gou12]. Then we can foliate any globally hyperbolic spacetimes $(\mathcal{M}, \bar{g})$ by a family of spacelike hypersurfaces $(\Sigma_t)_{t \in \mathbb{R}}$ in the following sense

$$\mathcal{M} = \bigcup_{t \in \mathbb{R}} \Sigma_t.$$

Roughly speaking, there is a regular smooth scalar field t on $\mathcal{M}$ in the sense that ∇t never vanishes such that each Σ_t is a level surface of t. Clearly, the regularity of t implies that $\Sigma_t \cap \Sigma_{t'} = \emptyset$ for any $t \neq t'$. Furthermore, we can use $\overline{\nabla} t$ to define time orientation for $(\mathcal{M}, \bar{g})$ and hence any other time like vector X is said to be

- *future pointing*: if $\bar{g}(X, \overline{\nabla} t)$ is negative or
- *past pointing*: if $\bar{g}(X, \overline{\nabla} t)$ is positive.

Suppose that Σ is spacelike, then we choose from (1.4) the following

$$\vec{n} = \frac{1}{\sqrt{-\overline{\nabla} t \cdot \overline{\nabla} t}} \overline{\nabla} t. \tag{1.5}$$

Clearly,

$$\bar{g}(\vec{n}, \overline{\nabla} t) = \bar{g}\left(\frac{\overline{\nabla} t}{(-\overline{\nabla} t \cdot \overline{\nabla} t)^{1/2}}, \overline{\nabla} t \right) = -(-\overline{\nabla} t \cdot \overline{\nabla} t)^{1/2} < 0.$$

In other words, $\vec{n}$ constructed before is future pointing normal.

1.3. *Adapted frames and coframes*

Given a spacetime $(\mathcal{M}, \bar{g}, \overline{\nabla})$ of dimension $n + 1$, we start by choosing an $(n + 1)$-foliation of the spacetime manifold $F_t : M \to \mathcal{M}$, $t \in \mathbb{R}$, for which each of the leaves $F_t(M)$ of the foliation, as a level set of the global time function t, is presumed spacelike. Recall from (1.5) that the future-directed normal unit vector $\vec{n}$ can be defined by using the gradient 1-form $\overline{\nabla} t$ which is nothing but

$$\vec{n} = N \overline{\nabla} t,$$

where N is the positive definite lapse function.

It is also convenient to choose a threading of the spacetime $\mathcal{M}$. The choice of a foliation and a threading together with a choice of coordinates $(x^1, ..., x^n)$ for M automatically induce local coordinates $(x^0 = t, x^1, ..., x^n)$ on $\mathcal{M}$. For a natural frame on M, we choose

$$\partial_i = \frac{\partial}{\partial x_i} \quad \forall i = \overline{1, n}.$$

Let us now denote by ∂_t the dual of $\overline{\nabla} t$, that is, $\langle \overline{\nabla} t, \partial_t \rangle = 1$. The vector ∂_t is usually called the *time vector*. Also, there is no risk to use both dt and $\overline{\nabla} t$ to denote the same gradient 1-form. Then we need to find the remaining ∂_0.

Assume that we have already had the frame $(\partial_0, \partial_1, ..., \partial_n)$. The dual coframe $(\theta^i)_{i=0}^n$ of $(\partial_i)_{i=0}^n$ is found to be such that

$$\theta^i = dx^i \mid \beta^i dt \quad \forall i = \overline{1, n},$$

while the 1-form θ^0 is nothing but dt. Here β^i are the components of the spacelike shift vector β which is the difference between $N\vec{n}$ and ∂_t. To be exact, the shift vector is chosen as follows

$$\beta = \partial_t - N\vec{n}. \tag{1.6}$$

Since

$$\langle \overline{\nabla} t, \beta \rangle = \langle \overline{\nabla} t, \partial_t - N\vec{n} \rangle = \langle \overline{\nabla} t, \partial_t \rangle - \langle \overline{\nabla} t, -N\vec{n} \rangle = 1 - 1 = 0.$$

The last thing we need to find is the vector ∂_0. To select ∂_0 suitable, we need to solve $\langle \partial_0, \theta^i \rangle = 0$ for all $i = \overline{1, n}$. Hence, the only possibility is that

$$\partial_0 = \partial_t - \beta^j \partial_j.$$

It is now easy to check that $\partial_0 = N\vec{n}$ in our case. It is worth noting that the presence of ∂_0 in the basis for $T\mathcal{M}$ is because the vector ∂_t is not always timelike since $\langle \partial_t, \partial_t \rangle < 0$ doest not hold in general as can be seen from the following computation

$$\langle \partial_t, \partial_t \rangle = \langle N\vec{n} + \beta, N\vec{n} + \beta \rangle = -N^2 + \langle \beta, \beta \rangle.$$

Note that in the previous identity, we have used the fact that β is tangential to the leave $F_t(M)$; hence $\langle \vec{n}, \beta \rangle = 0$.

By using the coframe $(\theta^i)_{i=0}^n$, we may locally rewrite the spacetime metric $\overline{g}$ in the form

$$\overline{g} = -N^2 \theta^0 \otimes \theta^0 + g_{ij} \theta^i \otimes \theta^j,$$

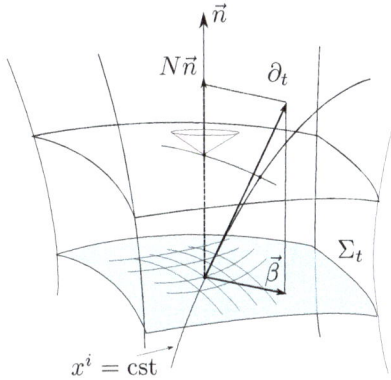

Fig. 2.   Shift vector β, lapse function N in the spacetime foliation.

where g_{ij} are the components of the spatial metric tensor g. In terms of the local coordinates $(x^0 = t, x^1, ..., x^n)$ on $\mathcal{M}$, we can write

$$\bar{g} = -N^2 dt \otimes dt + g_{ij}(dx^i + \beta^i dt) \otimes (dx^j + \beta^j dt). \tag{1.7}$$

We note that for each t, the term $g_{ij}(t)dx^i \otimes dx^j$ is the induced Riemannian metric on the leaf $F_t(M)$; hence we can write

$$(\bar{g}_{\alpha\beta}) = \begin{pmatrix} \bar{g}_{00} & \vdots \\ \cdots & (g_{ij}) \end{pmatrix}.$$

Besides, on each leaf $F_t(M)$, we have a tangent vector basis given by $(\partial_1, ..., \partial_n)$.

1.4. *Basics of submanifolds*

We now study the geometry of spacelike M of dimension n sitting in a spacetime $(\mathcal{M}, \bar{g}, \bar{\nabla})$ of dimension $n+1$ as an embedded submanifold. (M is being considered as Σ in previous subsections.) Hence, $\bar{g}$ induces a non-degenerate metric g on M called *the first fundamental form*, this is because M is spacelike.

Corresponding to g, there is the Levi–Civita connection ∇. Then the pair (g, ∇) and $(\bar{g}, \bar{\nabla})$ are related by

$$\bar{\nabla}_X Y = (\bar{\nabla}_X Y)^{/\!/} + (\bar{\nabla}_X Y)^\perp = \nabla_X Y + \mathbb{II}(X, Y). \tag{1.8}$$

The symmetric tensor $\mathbb{II}$ is called the *(vector-valued) second fundamental form* on M. (The first is the metric g, of course.) Since $\mathbb{II}$ is proportional to the normal $\vec{n}$ of M, we can write

$$\widetilde{K}(X,Y)\vec{n} = \mathbb{II}(X,Y) \tag{1.9}$$

and call $\widetilde{K}$ the *(scalar-valued) second fundamental form*. Sometimes, we also denote

$$K(X,Y) = \langle \mathbb{II}(X,Y), \vec{n} \rangle \tag{1.10}$$

and also call K the scalar-valued Second Fundamental Form; hence $\widetilde{K} = \langle \vec{n}, \vec{n} \rangle K$. In other words, K and $\widetilde{K}$ are different simply by the sign of $\langle \vec{n}, \vec{n} \rangle$. Also by (1.9), we obtain

$$\mathbb{II}(X,Y) = \langle \vec{n}, \vec{n} \rangle K(X,Y)\vec{n}.$$

Now, thanks to $\overline{\nabla}_Y \langle X, \vec{n} \rangle = 0$, we further obtain

$$K(X,Y) = \langle \mathbb{II}(X,Y), \vec{n} \rangle = \langle \overline{\nabla}_X Y, \vec{n} \rangle = -\langle \overline{\nabla}_X \vec{n}, Y \rangle.$$

1.5. *The Gauss and Codazzi equations*

In the next few sections, we shall see that Einstein constraint equations are simply consequences of the Gauss and Codazzi equations which will be considered in this subsection.

However, for the sake of clarity, we first derive several useful properties of the Riemann curvature tensor. First, recall from (1.1) that Rm acting on vector fields X, Y, and Z is

$$\overline{\mathrm{Rm}}(X,Y)Z = \overline{\nabla}_X \overline{\nabla}_Y Z - \overline{\nabla}_Y \overline{\nabla}_X Z - \overline{\nabla}_{[X,Y]} Z,$$

hence by interchanging X and Y we know that

$$\overline{\mathrm{Rm}}(Y,X)Z = \overline{\nabla}_Y \overline{\nabla}_X Z - \overline{\nabla}_X \overline{\nabla}_Y Z - \overline{\nabla}_{[Y,X]} Z = -\overline{\mathrm{Rm}}(X,Y)Z. \tag{1.11}$$

In other words, the Riemann curvature tensor is skew symmetric with respect to the first two components.

To prove that $\langle \mathrm{Rm}(X,Y)Z, W \rangle$ is also skew symmetric with respect to the last two components Z and W, we first show that

$$\mathbb{II}(Y,Z) = \mathbb{II}(Z,Y),$$

that is to say that the second fundamental form is symmetric. To see this, first observe by definition (1.8) that

$$\overline{\nabla}_Y Z = \nabla_Y Z + \mathbb{II}(Y,Z),$$
$$\overline{\nabla}_Z Y = \nabla_Z Y + \mathbb{II}(Z,Y).$$

Hence, using the torsion free property of connections $\overline{\nabla}$ and ∇, we obtain after subtracting the two equations

$$[Y, Z]_{\overline{\nabla}} = [Y, Z]_{\nabla} + \mathbb{I}(Y, Z) - \mathbb{I}(Z, Y).$$

From this we obtain the symmetry of $\mathbb{I}$. Next, to see why

$$\langle \mathrm{Rm}(X, Y)Z, W \rangle = \langle \mathrm{Rm}(X, Y)W, Z \rangle,$$

it suffices to show $\langle \mathrm{Rm}(X, Y)Z, Z \rangle = 0$ for any X, Y, Z.

We are now in a position to state the Gauss equation which tells us how $\overline{\mathrm{Rm}}(X, Y)Z$ and $\mathrm{Rm}(X, Y)Z$ are related to each other. We state it in the following proposition.

Proposition 1.1: *Suppose that (M, g, ∇) is a submanifold of $(\mathscr{M}, \overline{g}, \overline{\nabla})$. Then for any $X, Y, Z, W \in TM$, we have the following equation*

$$\langle \overline{\mathrm{Rm}}(X, Y)Z, W \rangle = \langle \mathrm{Rm}(X, Y)Z, W \rangle - \langle \mathbb{I}(X, Z), \mathbb{I}(Y, W) \rangle \tag{1.12}$$
$$+ \langle \mathbb{I}(Y, Z), \mathbb{I}(X, W) \rangle .$$

Proof: First, we express $\overline{\mathrm{Rm}}(X, Y)Z$ in terms of geometric quantities on (M, g, ∇). This can be done in detail as the following.

$$\begin{aligned}
\overline{\mathrm{Rm}}(X, Y)Z &= \overline{\nabla}_X \overline{\nabla}_Y Z - \overline{\nabla}_Y \overline{\nabla}_X Z - \overline{\nabla}_{[X,Y]} Z \\
&= \overline{\nabla}_X \left(\nabla_Y Z + \mathbb{I}(Y, Z) \right) - \overline{\nabla}_Y \left(\nabla_X Z + \mathbb{I}(X, Z) \right) \\
&\quad - \nabla_{[X,Y]} Z - \mathbb{I}([X, Y], Z) \\
&= \nabla_X \nabla_Y Z + \mathbb{I}(X, \nabla_Y Z) + \overline{\nabla}_X (\mathbb{I}(Y, Z)) \\
&\quad - \nabla_Y \nabla_X Z - \mathbb{I}(Y, \nabla_X Z) - \overline{\nabla}_Y (\mathbb{I}(X, Z)) \\
&\quad - \nabla_{[X,Y]} Z - \mathbb{I}([X, Y], Z) \\
&= \mathrm{Rm}(X, Y)Z + \mathbb{I}(X, \nabla_Y Z) - \mathbb{I}(Y, \nabla_X Z) - \mathbb{I}([X, Y], Z) \\
&\quad + \overline{\nabla}_X (\mathbb{I}(Y, Z)) - \overline{\nabla}_Y (\mathbb{I}(X, Z)).
\end{aligned}$$

The three terms involving $\mathbb{I}$ are perpendicular to W; hence by taking inner product with W with respect to the spacetime metric $\overline{g}$, we know that

$$\begin{aligned}
\langle \overline{\mathrm{Rm}}(X, Y)Z, W \rangle_{\overline{g}} \\
&= \langle \mathrm{Rm}(X, Y)Z, W \rangle_{\overline{g}} + \langle \overline{\nabla}_X (\mathbb{I}(Y, Z)), W \rangle_{\overline{g}} - \langle \overline{\nabla}_Y (\mathbb{I}(X, Z)), W \rangle_{\overline{g}} \\
&= \langle \mathrm{Rm}(X, Y)Z, W \rangle_g - \langle \mathbb{I}(Y, Z), \overline{\nabla}_X W \rangle_{\overline{g}} + \langle \mathbb{I}(X, Z), \overline{\nabla}_Y W \rangle_{\overline{g}} \\
&= \langle \mathrm{Rm}(X, Y)Z, W \rangle_g - \langle \mathbb{I}(Y, Z), \mathbb{I}(X, W) \rangle_{\overline{g}} + \langle \mathbb{I}(X, Z), \mathbb{I}(Y, W) \rangle_{\overline{g}}
\end{aligned}$$

as claimed. $\square$

In view of the Gauss equation (1.12), it is left with $\overline{\mathrm{Rm}}^{\perp}(X,Y)Z$ and this is the content of the Codazzi equation. Sometimes, we call it the Codazzi–Mainardi equation or the Ricci identity, which expresses the curvature of the normal bundle in terms of the second fundamental form.

Proposition 1.2: *Suppose that (M, g, ∇) is a submanifold of $(\mathcal{M}, \bar{g}, \overline{\nabla})$. Then for any $X, Y, Z, W \in TM$, we have the following equation*

$$\overline{\mathrm{Rm}}^{\perp}(X,Y)Z = (\overline{\nabla}_X \mathbb{II})(Y,Z) - (\overline{\nabla}_Y \mathbb{II})(X,Z) \tag{1.13}$$

where the derivatives $\overline{\nabla}_X \mathbb{II}$ is understood as follows

$$(\overline{\nabla}_U \mathbb{II})(V,W) = \overline{\nabla}_U^{\perp}(\mathbb{II}(V,W)) - \mathbb{II}(\nabla_U V, W) - \mathbb{II}(V, \nabla_U W)$$

for any tangent vector fields U, V, W in TM.

Proof: To see this, we start with

$$\begin{aligned} \overline{\mathrm{Rm}}(X,Y)Z = \mathrm{Rm}(X,Y)Z + \mathbb{II}(X, \nabla_Y Z) - \mathbb{II}(Y, \nabla_X Z) - \mathbb{II}([X,Y], Z) \\ + \overline{\nabla}_X(\mathbb{II}(Y,Z)) - \overline{\nabla}_Y(\mathbb{II}(X,Z)) \end{aligned}$$

which we have already derived before. Then we look at the normal part of the both sides to get

$$\begin{aligned} \overline{\mathrm{Rm}}^{\perp}(X,Y)Z = \mathbb{II}(X, \nabla_Y Z) - \mathbb{II}(Y, \nabla_X Z) - \mathbb{II}([X,Y], Z) \\ + \overline{\nabla}_X^{\perp}(\mathbb{II}(Y,Z)) - \overline{\nabla}_Y^{\perp}(\mathbb{II}(X,Z)). \end{aligned}$$

Using the fact $[X,Y] = \nabla_X Y - \nabla_Y X$ to get

$$\mathbb{II}([X,Y], Z) = \mathbb{II}(\nabla_X Y, Z) - \mathbb{II}(\nabla_Y X, Z).$$

From this we eventually get

$$\begin{aligned} \overline{\mathrm{Rm}}^{\perp}(X,Y)Z &= \mathbb{II}(X, \nabla_Y Z) - \mathbb{II}(Y, \nabla_X Z) - \mathbb{II}(\nabla_X Y, Z) + \mathbb{II}(\nabla_Y X, Z) \\ &\quad + \overline{\nabla}_X^{\perp}(\mathbb{II}(Y,Z)) - \overline{\nabla}_Y^{\perp}(\mathbb{II}(X,Z)) \\ &= (\overline{\nabla}_X \mathbb{II})(Y,Z) - (\overline{\nabla}_Y \mathbb{II})(X,Z) \end{aligned}$$

as claimed.   □

2. The Einstein Equations in General Relativity

2.1. *The Einstein equations*

As always, we assume throughout this section that $(\mathcal{M}, \bar{g})$ is a Lorentzian manifold of dimension $n + 1$. In this notes, we do not want to address the question: *How could Einstein come up with the equation named after him?* For the sake of simplicity, we adopt the Einstein equation as an axiom equation. However, like other field equations in physics, the original Einstein equations can be derived from an action through the principle of least action. More precise, the action

$$\int_{\mathcal{M}} (\text{Scal}_{\bar{g}} + 2\mathscr{L}_{\bar{g}}) \, d\text{vol}_{\bar{g}} \qquad (2.1)$$

is required to be stable under compact perturbations of the metric $\bar{g}$ where $\mathscr{L}_{\bar{g}}$ is the Lagrangian associated with non-gravitational fields. When $\mathscr{L}_{\bar{g}} \equiv 0$, the action (2.1) is known as the Einstein–Hilbert action.

For each compactly supported 2-tensor $\bar{h}$, by direct computing[a], we obtain the variation of $\int_{\mathcal{M}} \text{Scal}_{\bar{g}} \, d\text{vol}_{\bar{g}}$ with respect to $\bar{g}$ as follows

$$\frac{d}{dt}\Big|_{t=0} \int_{\mathcal{M}} \text{Scal}_{\bar{g}+t\bar{h}} \, d\text{vol}_{\bar{g}+t\bar{h}} = -\int_{\mathcal{M}} \bar{h}\left(\text{Ric}_{\bar{g}} - \frac{1}{2}\text{Scal}_{\bar{g}}\,\bar{g}\right) d\text{vol}_{\bar{g}}. \qquad (2.2)$$

---

[a]Perhaps, it would be easier to derive Eq. (2.4) using "delta" variation instead of using (2.2). Indeed, there holds

$$0 = \delta \int_{\mathcal{M}} \text{Scal}_{\bar{g}} \, d\text{vol}_{\bar{g}}$$

$$= \int_{\mathcal{M}} \left(\frac{\delta[\sqrt{|\det(\bar{g})|}\,\text{Scal}_{\bar{g}}]}{\delta\bar{g}^{ij}}\right) \delta\bar{g}^{ij} \, dx$$

$$= \int_{\mathcal{M}} \left(\frac{\delta\sqrt{|\det(\bar{g})|}}{\delta\bar{g}^{ij}} \frac{\text{Scal}_{\bar{g}}}{\sqrt{|\det(\bar{g})|}} + \frac{\delta\,\text{Scal}_{\bar{g}}}{\delta\bar{g}^{ij}}\right) \delta\bar{g}^{ij} \, d\text{vol}_{\bar{g}}.$$

Then it is easy to obtain the left hand side of Eq. (2.4) by noticing that

$$\frac{\delta\sqrt{|\det(\bar{g})|}}{\delta\bar{g}^{ij}} = -\frac{1}{2}\bar{g}\sqrt{|\det(\bar{g})|}, \qquad \frac{\delta\,\text{Scal}_{\bar{g}}}{\delta\bar{g}} = \text{Ric}_{\bar{g}}.$$

Regarding to the Lagrangian $\mathscr{L}_{\overline{g}}$ [b], we obtain the following formula

$$\frac{d}{dt}\bigg|_{t=0} \int_{\mathscr{M}} \mathscr{L}_{\overline{g}+t\overline{h}}\, \mathrm{dvol}_{\overline{g}+t\overline{h}} = \int_{\mathscr{M}} \left((D_{\overline{g}}\mathscr{L}_{\overline{g}})(h) + \mathscr{L}_{\overline{g}} \cdot \frac{1}{2}\operatorname{tr}_{\overline{g}}(\overline{h}) \right) \mathrm{dvol}_{\overline{g}}. \quad (2.3)$$

Hence, by the variation of (2.1) with respect to $\overline{g}$, we obtain the field equation

$$\mathrm{Ric}_{\overline{g}} - \frac{1}{2}\mathrm{Scal}_{\overline{g}}\,\overline{g} = T \qquad\qquad (2.4)$$

where the left hand side of Eq. (2.4) is known as the Einstein tensor which will be denoted by $\mathrm{Eins}_{\overline{g}}$ and the tensor T, which comes from the variation of $\int_{\mathscr{M}} \mathscr{L}_{\overline{g}}\, \mathrm{dvol}_{\overline{g}}$ with respect to $\overline{g}$, is known as the stress-energy tensor. For particular Lagrangian $\mathscr{L}_{\overline{g}}$, we can easily calculate the stress-energy tensor T as we shall see later when scalar fields are included in our spacetime.

By taking the trace of the Einstein tensor, we further arrive at

$$\operatorname{tr}(\mathrm{Eins}_{\overline{g}}) = \operatorname{tr}\left(\mathrm{Ric}_{\overline{g}} - \frac{1}{2}\mathrm{Scal}_{\overline{g}}\,\overline{g} \right) = \frac{1-n}{2}\mathrm{Scal}_{\overline{g}}.$$

A vacuum space time is a Lorentzian manifold $(\mathscr{M}, \overline{g})$ that satisfies Eq. (2.4) with vanishing stress-energy tensor $T = 0$, thanks to $\mathscr{L}_{\overline{g}} \equiv 0$ and (2.3). In this case, we can take the trace of Eq. (2.4) to find $\mathrm{Scal}_{\overline{g}} = 0$ and obtain the vacuum Einstein equations $\mathrm{Ric}_{\overline{g}} = 0$. We conclude this subsection by noting that the fully Einstein equations (with cosmological constant Λ) may be written in the following form

$$\mathrm{Eins}_{\overline{g}} + \Lambda\overline{g} = T$$

where Λ is constant called the cosmological constant. The cosmological constant term was originally introduced by Einstein to allow for a static universe. It is clear to see that the cosmological term could be absorbed into the stress-energy tensor T as dark energy.

---

[b]To derive (2.3), we can again use "delta" variation. Indeed, there holds

$$\delta \int_{\mathscr{M}} \mathscr{L}_{\overline{g}}\, \mathrm{dvol}_{\overline{g}} = \int_{\mathscr{M}} \left(\frac{\delta\left[\sqrt{|\det(\overline{g})|}\mathscr{L}_{\overline{g}}\right]}{\delta\overline{g}^{ij}} \right)\delta\overline{g}^{ij}\, \mathrm{d}x$$

$$= \int_{\mathscr{M}} \left(\frac{\delta\mathscr{L}_{\overline{g}}}{\delta\overline{g}^{ij}} + \frac{\mathscr{L}_{\overline{g}}}{\sqrt{|\det(\overline{g})|}}\frac{\delta\left[\sqrt{|\det(\overline{g})|}\right]}{\delta\overline{g}^{ij}} \right)\delta\overline{g}^{ij}\, \mathrm{dvol}_{\overline{g}}$$

$$= \int_{\mathscr{M}} \left(\frac{\delta\mathscr{L}_{\overline{g}}}{\delta\overline{g}^{ij}} - \frac{1}{2}\overline{g}_{ij}\mathscr{L}_{\overline{g}} \right)\delta\overline{g}^{ij}\, \mathrm{dvol}_{\overline{g}}.$$

Thus, we also obtain

$$T_{ij} = -2\frac{\delta\mathscr{L}_{\overline{g}}}{\delta\overline{g}^{ij}} + \overline{g}_{ij}\mathscr{L}_{\overline{g}}.$$

2.2. The Einstein equations with real scalar fields

As can be seen from Eq. (2.4), the Einstein equations involve the so-called stress-energy tensor T representing the density of all the energies, momentum, and stresses of the sources. On a macroscopic scale, one can couple gravity to either field sources or matter sources such as electro-magnetic fields, Yang-Mills sources, and scalar fields. While the latter are more phenomenological, the former, which are deduced directly from special relativity, are one of the simplest non-vacuum models which are the core of studies in recent years. In addition, interest in those models stems partly from recent attempts to use such models to study the observed acceleration of the expansion of the universe.

From now on, we only focus on the Einstein equations equipped with scalar fields. In modern cosmology, one can introduce on the spacetime $(\mathcal{M}, \overline{g})$ a real scalar field $\overline{\psi}$ with potential U as a smooth function of $\overline{\psi}$. A particular Einstein field theory is specified by the choice of an action principle, for example, with the following Lagrangian

$$\mathscr{L}_{\overline{g}} = -\frac{1}{2}|\nabla\overline{\psi}|_{\overline{g}}^2 - U(\overline{\psi}). \tag{2.5}$$

This choice of action principle also includes the well-known massive or massless Klein-Gordon field theory where $U(\overline{\psi}) = m^2\overline{\psi}^2/2$. Thanks to (2.3), by a fairy standard computation[c], one can deduce that

$$T_{\alpha\beta} = \nabla_\alpha\overline{\psi}\nabla_\beta\overline{\psi} - \frac{1}{2}\overline{g}_{\alpha\beta}\nabla_\mu\overline{\psi}\nabla^\mu\overline{\psi} - \overline{g}_{\alpha\beta}U(\overline{\psi}). \tag{2.6}$$

It is worth noticing that a cosmological constant Λ can be considered as a particular scalar field with potential Λ. For this reason, we do not consider the cosmological term in our Einstein equation.

---

[c]To derive (2.6), we can again use "delta" variation. Recall the following formula

$$T_{ij} = -2\frac{\delta\mathscr{L}_{\overline{g}}}{\delta\overline{g}^{ij}} + \overline{g}_{ij}\mathscr{L}_{\overline{g}}.$$

Hence

$$\begin{aligned} T_{ij} &= -2\frac{\delta}{\delta\overline{g}^{ij}}\left(-\frac{1}{2}\overline{g}^{ij}\nabla_i\overline{\psi}\nabla_j\overline{\psi} - U(\overline{\psi})\right) + \overline{g}_{ij}\left(-\frac{1}{2}|\nabla\overline{\psi}|_{\overline{g}}^2 - U(\overline{\psi})\right) \\ &= \nabla_i\overline{\psi}\nabla_j\overline{\psi} - \frac{1}{2}\overline{g}_{ij}|\nabla\overline{\psi}|_{\overline{g}}^2 - \overline{g}_{ij}U(\overline{\psi}). \end{aligned}$$

2.3. Why do the Einstein equations describe the propagation of wavelike phenomena?

In order to formulate the initial value problem for the Einstein equations as nonlinear wave equations, we first express the Einstein equations in terms of a partial differential equation along with some gauge condition; hence the Einstein equations basically describe the propagation of wave.

We suppose that $(\mathscr{M}, \overline{g})$ is a Lorentzian manifold of the dimension $n+1$. In a coordinate system that will be fixed from now on, we have

$$\overline{\Gamma}^k_{ij} = \frac{1}{2}\overline{g}^{km}(\overline{g}_{im,j} + \overline{g}_{jm,i} - \overline{g}_{ij,m}) \tag{2.7}$$

as Christoffel symbols for the spacetime metric $\overline{g}$. For the sake of simplicity, by "lower orders" we mean terms consisting of either no derivative or first order derivative of the spacetime metric $\overline{g}$. As such, terms consisting of derivatives of order higher than two will be called high order terms.

We now denote by $\Box_{\overline{g}}$ the wave operator (or the d'Alembertian) evaluated with respect to the indicated metric $\overline{g}$. Mathematically, $\Box_{\overline{g}}$ is defined to be

$$\Box_{\overline{g}} = \overline{\nabla}^i\overline{\nabla}_i = \overline{g}^{ij}\overline{\nabla}_i\overline{\nabla}_j. \tag{2.8}$$

Then, we introduce the following notation

$$\lambda^\alpha = \Box_{\overline{g}}x^\alpha$$

where x^α are coordinate functions. Using (2.8) and thanks to the rule $\overline{\nabla}_i\partial_j = \Gamma^k_{ij}\partial_k$, we obtain

$$\lambda^\alpha = \overline{g}^{ij}\overline{\nabla}_i\overline{\nabla}_jx^\alpha = \overline{g}^{ij}(\overline{\Gamma}^k_{ij}x^\alpha_{,k}) = \overline{g}^{ij}\overline{\Gamma}^\alpha_{ij}.$$

Taking another derivative, it is obvious to see that

$$(\overline{g}_{\alpha i}\lambda^\alpha_{,j} + \overline{g}_{\alpha j}\lambda^\alpha_{,i}) = \overline{g}_{\alpha i}\overline{g}^{km}\overline{\Gamma}^\alpha_{km,j} + \overline{g}_{\alpha j}\overline{g}^{km}\overline{\Gamma}^\alpha_{km,i}.$$

Using (2.7), we can write

$$\overline{\Gamma}^\alpha_{km,j} = \frac{1}{2}\overline{g}^{\alpha p}(\overline{g}_{kp,mj} + \overline{g}_{mp,kj} - \overline{g}_{km,pj}) + \text{lower orders}$$

and

$$\overline{\Gamma}^\alpha_{km,i} = \frac{1}{2}\overline{g}^{\alpha q}(\overline{g}_{kq,mi} + \overline{g}_{mq,ki} - \overline{g}_{km,qi}) + \text{lower orders}.$$

Therefore, using the Kronecker symbol δ, for example $\bar{g}_{\alpha i}\bar{g}^{\alpha p} = \delta^p_i$, a further calculation shows that

$$(\bar{g}_{\alpha i}\lambda^\alpha_{,j} + \bar{g}_{\alpha j}\lambda^\alpha_{,i}) = \frac{1}{2}\bar{g}_{\alpha i}\bar{g}^{km}\bar{g}^{\alpha p}(\bar{g}_{kp,mj} + \bar{g}_{mp,kj} - \bar{g}_{km,pj})$$

$$+ \frac{1}{2}\bar{g}_{\alpha j}\bar{g}^{km}\bar{g}^{\alpha q}(\bar{g}_{kq,mi} + \bar{g}_{mq,ki} - \bar{g}_{km,qi}) + \text{lower orders}$$

$$= \frac{1}{2}\delta^p_i\bar{g}^{km}(\bar{g}_{kp,mj} + \bar{g}_{mp,kj} - \bar{g}_{km,pj})$$

$$+ \frac{1}{2}\delta^q_j\bar{g}^{km}(\bar{g}_{kq,mi} + \bar{g}_{mq,ki} - \bar{g}_{km,qi}) + \text{lower orders}$$

$$= \frac{1}{2}\bar{g}^{km}(\bar{g}_{ki,mj} + \bar{g}_{mi,kj} - \bar{g}_{km,ij})$$

$$+ \frac{1}{2}\bar{g}^{km}(\bar{g}_{kj,mi} + \bar{g}_{mj,ki} - \bar{g}_{km,ij}) + \text{lower orders}$$

$$= \bar{g}^{km}(\bar{g}_{ki,mj} + \bar{g}_{mi,kj} - \bar{g}_{km,ij}) + \text{lower orders}.$$

Thus, we have just shown that

$$\bar{g}^{km}(\bar{g}_{ki,mj} + \bar{g}_{mi,kj} - \bar{g}_{km,ij}) = (\bar{g}_{\alpha i}\lambda^\alpha_{,j} + \bar{g}_{\alpha j}\lambda^\alpha_{,i}) + \text{lower orders}. \quad (2.9)$$

We now have the components of the Ricci curvature $\mathrm{Ric}_{\bar{g}}$ of $\bar{g}$ given by

$$\overline{\mathrm{Ric}}_{ij} = \bar{\Gamma}^k_{ij,k} - \bar{\Gamma}^k_{ik,j} + \bar{\Gamma}^k_{kl}\bar{\Gamma}^l_{ij} - \bar{\Gamma}^k_{jl}\bar{\Gamma}^l_{ik}.$$

A simple observation tells us that the difference $\bar{\Gamma}^k_{kl}\bar{\Gamma}^l_{ij} - \bar{\Gamma}^k_{jl}\bar{\Gamma}^l_{ik}$ consists of lower order terms only. Therefore, we can write

$$\overline{\mathrm{Ric}}_{ij} = \bar{\Gamma}^k_{ij,k} - \bar{\Gamma}^k_{ik,j} + \text{lower orders}.$$

Moreover, using (2.7) we calculate the difference $\bar{\Gamma}^k_{ij,k} - \bar{\Gamma}^k_{ik,j}$ as follows

$$\bar{\Gamma}^k_{ij,k} - \bar{\Gamma}^k_{ik,j} = \frac{1}{2}\bar{g}^{km}(\bar{g}_{im,kj} + \bar{g}_{jm,ki} - \bar{g}_{ij,km})$$

$$- \frac{1}{2}\bar{g}^{km}(\bar{g}_{im,kj} + \bar{g}_{km,ij} - \bar{g}_{ik,mj}) + \text{lower orders}$$

$$= \frac{1}{2}\bar{g}^{km}(\bar{g}_{jm,ki} - \bar{g}_{ij,km} - \bar{g}_{km,ij} + \bar{g}_{ik,mj}) + \text{lower orders}.$$

In other words,

$$\overline{\mathrm{Ric}}_{ij} = \frac{1}{2}\bar{g}^{km}(\bar{g}_{jm,ki} - \bar{g}_{ij,km} - \bar{g}_{km,ij} + \bar{g}_{ik,mj}) + \text{lower orders}. \quad (2.10)$$

Combining (2.10) and our calculation for $\bar{g}_{\alpha i}\lambda^\alpha_{,j} + \bar{g}_{\alpha j}\lambda^\alpha_{,i}$ in (2.9) above, we find that

$$\overline{\mathrm{Ric}}_{ij} = \frac{1}{2}(\bar{g}_{\alpha i}\lambda^\alpha_{,j} + \bar{g}_{\alpha j}\lambda^\alpha_{,i}) - \frac{1}{2}\bar{g}^{km}\bar{g}_{ij,km} + \text{lower orders}. \quad (2.11)$$

In view of the Einstein equation (2.4), we obtain

$$\overline{\mathrm{Ric}}_{ij} = T_{ij} + \frac{1}{2}\overline{g}_{ij}\,\mathrm{Scal}_{\overline{g}},$$

which then helps us to conclude

$$-\frac{1}{2}\overline{g}^{km}\overline{g}_{ij,km} = \underbrace{-\frac{1}{2}(\overline{g}_{\alpha i}\lambda^{\alpha}_{,j} + \overline{g}_{\alpha j}\lambda^{\alpha}_{,i}) + T_{ij} + \frac{1}{2}\overline{g}_{ij}\,\mathrm{Scal}_{\overline{g}} + \text{lower orders}}_{\text{lower orders}}.$$

In order to get rid of high order terms on the right hand side of the preceding equation, i.e. the term $\overline{g}_{\alpha i}\lambda^{\alpha}_{,j} + \overline{g}_{\alpha j}\lambda^{\alpha}_{,i}$, it is usually to assume that

$$\lambda^{\alpha} = 0, \quad \forall \alpha = \overline{0, n}.$$

This condition is preferred to the so-called Lorentz-harmonic coordinate gauge. In the literature, this belongs to a set of a few condition that one can solve the Einstein equations.

Thus, we have shown that the Einstein equations for $\overline{g}$ in the Lorentz-harmonic coordinate gauge are nothing but wave equations for $\overline{g}$. We conclude this subsection by observing that on any Lorentzian manifold the Lorentz-harmonic coordinate gauge is fulfilled by simply solving the Cauchy problem for some linear wave equations. In addition, the solvability of the transformed wave-like system as shown above is well-understood by the seminal work of Leray.

3. The Cauchy Problem for the Einstein Equations

Since the Einstein equations are geometric equations, one can expect solutions of the Einstein equations verifying the causality principle, that is, the relativistic spacetime future cannot influence the past. Based on two works by Leray [Ler53] and Geroch [Ger70], we know that the globally hyperbolic spacetime $(\mathcal{M}, \overline{g})$ are topological products $M \times \mathbb{R}$ with each $M \times \{t\}$ intersected once by each inextensible timelike curve. In view of our definition above, such a submanifold $M \times \{t\}$ is a Cauchy surface.

In order to formulate an appropriate Cauchy problem for the Einstein equations, we assume that the globally hyperbolic spacetime $(\mathcal{M}, \overline{g})$ admits M as a Cauchy surface. We let g be the Riemannian metric on M induced by $\overline{g}$. Having such a Cauchy surface, we let $\vec{n}$ be the future pointing timelike unit normal vector to M. We also let K be the extrinsic curvature of M computed with respect to $\vec{n}$. We are now able to formulate the Cauchy problem for the Einstein equations intrinsically through the following definition,

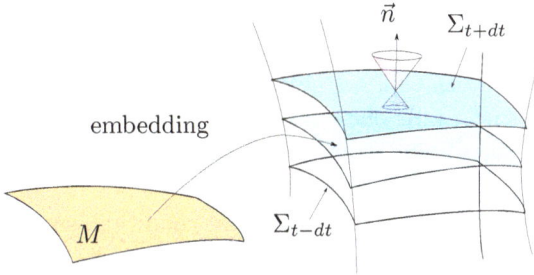

Fig. 3.   The Cauchy problem for the Einstein equations.

see [CBru09, Chapter VI].

Definition 3.1:

(1) An initial data set is a triplet (M, g, K) where M is an n-dimensional smooth manifold, g is a properly Riemannian metric on M and K a symmetric 2-tensor.

(2) A Cauchy development of an initial data set is a spacetime $(\mathcal{M}, \bar{g})$ such that there exists an embedding $\iota : M \to \mathcal{M}$ enjoying the following properties:

   (a) The metric g is the pullback of $\bar{g}$ by ι, that is, $g = \iota^* \bar{g}$. In other words, if M is identified with its image $\iota(M) = M_0$ in $\mathcal{M}$, then g is the metric induced by $\bar{g}$ on M_0.

   (b) $\iota(K)$ is the second fundamental form of $\iota(M)$ as submanifold of $(\mathcal{M}, \bar{g})$.

(3) A Cauchy development $(\mathcal{M}, \bar{g})$ of (M, g, K) is called a Einsteinian development if the metric $\bar{g}$ satisfies the Einstein equations on $\mathcal{M}$.

We now suppose that $(\mathcal{M}, \bar{g})$ is a Cauchy development of (M, g, K). If it is also true that every other Cauchy development of (M, g, K) can be isometrically embedded in $\mathcal{M}$, we say $\mathcal{M}$ is called the maximal development of (M, g, K). It is easy to see that a maximal development is unique up to isomorphism.

As we shall see later in Proposition 4.1, in order to generate a Cauchy development, the initial data (M, g, K) for the Einstein equations cannot be arbitrary, they must satisfy some conditions, this is because data on (M, g, K) must satisfy certain conditions when sitting in the ambient space. To be exact, in view of the Gauss and Codazzi equations, those conditions can be rewritten in a form consisting two equations known as the Hamiltonian and momentum constraints, see Eq. (4.1) and Eq. (4.2) below.

Conversely, we wish that for given initial data verifying the constraint equations (4.1) and (4.2), one can generate a corresponding Cauchy development the for such given initial data. The following theorem due to Y. Choquet-Bruhat and R. Geroch shows that this is possible in the sense that Eqs. (4.1) and (4.2) also give us a sufficient condition, see [CBG69].

Theorem 3.2: *Given smooth initial data (M, g, K) satisfying the constraint equations, there exists a smooth, maximal, globally hyperbolic Cauchy development of the initial data.*

Consequently, there is a strong connection between the globally hyperbolic Cauchy development of the initial data and the Einstein constraint equations. However, there is no one-to-one correspondence between solutions of the Einstein constraint equations and their globally hyperbolic maximal Cauchy developments in the sense that two distinct solutions of the Einstein constraint equations may generate isometric globally hyperbolic maximal Cauchy developments. Nevertheless, as a first step, it is important to understand solutions of the constraint equations.

4. Constraints and Evolutions

As we have already mentioned in the previous section, the initial data (M, g, K) for the Einstein equations cannot be arbitrary, they must satisfy some conditions. The purpose of this section is to prove Proposition 4.1 below which tells us how the initial data (M, g, K) are related to each others. A proof for Proposition 4.1 simply makes use of the Gauss and Codazzi equations, see Propositions 1.1 and 1.2 in Subsection 1.5.

Proposition 4.1: *The initial data (M, g, K) satisfies the following constraints:*

$$\text{Scal}_g - |K|_g^2 + (\text{tr}_g K)^2 = 2\rho \tag{4.1}$$

and

$$\text{div}_g K - d(\text{tr}_g K) = J, \tag{4.2}$$

where ρ is a scalar and J a vector on M determined by the projection on M and the normal to M, when embedded in a spacetime $(\mathcal{M}, \overline{g})$, of the tensor T of the sources.

In the literature, Eq. (4.1) is known as the Hamiltonian constraint while Eq. (4.2) is called the momentum constraint. We spend this section to prove Proposition 4.1.

4.1. *Derivation of constraint equations: The Hamiltonian constraint*

This subsection is devoted to prove (4.1) while (4.2) will be considered in the next subsection. Before we start, let us recall the Einstein equation without the cosmological constant Λ, that is

$$\operatorname{Ric}_{\overline{g}} - \frac{1}{2}\overline{g}\operatorname{Scal}_{\overline{g}} = T.$$

The above equation is understood over the Lorentzian manifold $(\mathcal{M}, \overline{g})$ of the dimension $n+1$. As always, (M, g) is being considered as a submanifold of $\mathcal{M}$ of the dimension n. Recall from (1.8), there holds

$$\nabla_X Y = \overline{\nabla}_X^{/\!/} Y, \quad \mathbb{I}(X,Y) = \overline{\nabla}_X^{\perp} Y$$

and $\overline{\nabla}_X Y = \nabla_X Y + \mathbb{I}(X,Y)$. Also, let us recall the Riemann curvature tensor $\overline{\mathrm{Rm}}$ given by

$$\overline{\mathrm{Rm}}(X,Y)Z = \overline{\nabla}_X \overline{\nabla}_Y Z - \overline{\nabla}_Y \overline{\nabla}_X Z - \overline{\nabla}_{[X,Y]} Z.$$

Another ingredient in the proof of (4.1) is the Gauss equation (1.12) is given by

$$\begin{aligned}
\langle \overline{\mathrm{Rm}}(X,Y)Z, W \rangle = {}& \langle \mathrm{Rm}\,(X,Y)Z), W \rangle - \langle \mathbb{I}(X,Z), \mathbb{I}(Y,W) \rangle \\
& + \langle \mathbb{I}(Y,Z), \mathbb{I}(X,W) \rangle.
\end{aligned}$$

We now let $e_1, ..., e_n$ be a local orthonormal frame field for M and we fix $e_0 = \vec{n}$. Using the Gauss equation (1.12) applied to $(X, Y, Z, W) = (e_i, e_j, e_i, e_j)$, we arrive at

$$\begin{aligned}
- \langle \overline{\mathrm{Rm}}\,(e_i, e_j)\, e_i, e_j \rangle = {}& - \langle \mathrm{Rm}(e_i, e_j)e_i, e_j \rangle + \langle \mathbb{I}(e_i, e_i), \mathbb{I}(e_j, e_j) \rangle \\
& - \langle \mathbb{I}(e_i, e_j), \mathbb{I}(e_i, e_j) \rangle,
\end{aligned}$$

which, after summing all over i and j from 1 to n, yields

$$\begin{aligned}
- \sum_{i,j=1}^{n} \langle \overline{\mathrm{Rm}}\,(e_i, e_j)\, e_i, e_j \rangle = {}& - \sum_{i,j=1}^{n} \langle \mathrm{Rm}(e_i, e_j)e_i, e_j \rangle \\
& + \sum_{i,j=1}^{n} \langle \mathbb{I}(e_i, e_i), \mathbb{I}(e_j, e_j) \rangle - \sum_{i,j=1}^{n} \langle \mathbb{I}(e_i, e_j), \mathbb{I}(e_i, e_j) \rangle.
\end{aligned} \tag{4.3}$$

By the definition of the scalar curvature, we have

$$\mathrm{Scal}_g = \sum_{j=1}^{n} \langle \mathrm{Ric}(e_j), e_j \rangle = - \sum_{i,j=1}^{n} \langle \mathrm{Rm}(e_i, e_j)e_i, e_j \rangle,$$

where $\mathrm{Ric}(u) = \sum_j \mathrm{Rm}(u, e_j)e_j$. Notice that the Ricci tensor is skew symmetric, thanks to (1.11). Besides, by the definition of the (scalar-valued) second fundamental form $K(\cdot, \cdot)$ with respect to the unit normal vector e_0, i.e.

$$K(X, Y) = \langle \mathrm{II}(X, Y), e_0 \rangle.$$

We recall that the corresponding mean curvature $\mathrm{tr}_g K$ of M, denoted by τ, is nothing but the trace of the second fundamental form K; hence τ can be calculated using

$$\tau = \mathrm{tr}_g(K) = \sum_{i=1}^{n} K(e_i, e_i).$$

In the case of a hypersurface, by using (1.10), there holds

$$\mathrm{II}(X, Y) = \langle e_0, e_0 \rangle K(X, Y)e_0, \tag{4.4}$$

regardless of the sign of $\langle e_0, e_0 \rangle$. Therefore, applying (4.4) to $(X, Y) = (e_i, e_i)$ to get

$$\sum_{i,j=1}^{n} \langle \mathrm{II}(e_i, e_i), \mathrm{II}(e_j, e_j) \rangle = \sum_{i,j=1}^{n} \langle \langle e_0, e_0 \rangle K(e_i, e_i)e_0, \langle e_0, e_0 \rangle K(e_j, e_j)e_0 \rangle$$

$$= \langle e_0, e_0 \rangle \sum_{i,j=1}^{n} K(e_i, e_i)K(e_j, e_j)$$

$$= - \left(\sum_{i=1}^{n} K(e_i, e_i) \right) \left(\sum_{j=1}^{n} K(e_j, e_j) \right)$$

$$= -(\mathrm{tr}_g K)^2. \tag{4.5}$$

For the remaining term, we again apply (4.4) to $(X, Y) = (e_i, e_j)$ to obtain

$$\sum_{i,j=1}^{n} \langle \mathbb{II}(e_i, e_j), \mathbb{II}(e_i, e_j) \rangle = \sum_{i,j=1}^{n} \langle \langle e_0, e_0 \rangle K(e_i, e_j) e_0, \langle e_0, e_0 \rangle K(e_i, e_j) e_0 \rangle$$

$$= \langle e_0, e_0 \rangle \sum_{i,j=1}^{n} K(e_i, e_j) K(e_i, e_j)$$

$$= - \sum_{i,j=1}^{n} K(e_i, e_j)^2$$

$$= -|K|_g^2.$$

$$(4.6)$$

Combining (4.3), (4.5), and (4.6), we have just shown that

$$- \sum_{i,j=1}^{n} \langle \overline{\mathrm{Rm}}\,(e_i, e_j)\, e_i, e_j \rangle = \mathrm{Scal}_g - |K|_g^2 + (\mathrm{tr}_g K)^2. \qquad (4.7)$$

Comparing (4.1) and (4.7), it remains to estimate the left hand side of (4.7). By definition of the Ricci curvature and the fact $\langle e_0, e_0 \rangle = -1$, we have

$$\mathrm{Ric}_{\overline{g}}(e_i, e_i) = -\langle \overline{\mathrm{Rm}}\,(e_0, e_i)\, e_i, e_0 \rangle + \sum_{j=1}^{n} \langle \overline{\mathrm{Rm}}\,(e_j, e_i)\, e_i, e_j \rangle$$

$$= \langle \overline{\mathrm{Rm}}\,(e_i, e_0)\, e_i, e_0 \rangle - \sum_{j=1}^{n} \langle \overline{\mathrm{Rm}}\,(e_i, e_j)\, e_i, e_j \rangle.$$

Therefore, thanks to $\langle \mathrm{Rm}(e_0, e_0) e_0, e_0 \rangle = 0$, we get

$$- \sum_{i,j=1}^{n} \langle \overline{\mathrm{Rm}}\,(e_i, e_j)\, e_i, e_j \rangle = \sum_{i=1}^{n} \left(\mathrm{Ric}_{\overline{g}}(e_i, e_i) - \langle \overline{\mathrm{Rm}}\,(e_i, e_0)\, e_i, e_0 \rangle \right)$$

$$= \sum_{i=1}^{n} \left(\mathrm{Ric}_{\overline{g}}(e_i, e_i) + \langle \overline{\mathrm{Rm}}\,(e_0, e_i)\, e_i, e_0 \rangle \right)$$

$$= \mathrm{Ric}_{\overline{g}}(e_0, e_0) + \sum_{i=1}^{n} \mathrm{Ric}_{\overline{g}}(e_i, e_i)$$

$$= 2\,\mathrm{Ric}_{\overline{g}}(e_0, e_0) + \left(-\mathrm{Ric}_{\overline{g}}(e_0, e_0) + \sum_{i=1}^{n} \mathrm{Ric}_{\overline{g}}(e_i, e_i) \right)$$

$$= 2\,\mathrm{Ric}_{\overline{g}}(e_0, e_0) + \mathrm{Scal}_{\overline{g}}.$$

Since $\mathrm{Ric}_{\overline{g}} - \frac{1}{2}\overline{g}\,\mathrm{Scal}_{\overline{g}} = T$, we obtain

$$\underbrace{\mathrm{Ric}_{\overline{g}}(e_0, e_0) - \frac{1}{2}\overline{g}(e_0, e_0)\,\mathrm{Scal}_{\overline{g}}}_{\mathrm{Ric}_{\overline{g}}(e_0, e_0) + \frac{1}{2}\,\mathrm{Scal}_{\overline{g}}} = T(e_0, e_0).$$

Thus, we have proved that

$$\mathrm{Scal}_g - |K|_g^2 + (\mathrm{tr}_g K)^2 = 2\rho,$$

known as the Hamitonian constraint equation. Note that here we denote $\rho = T(e_0, e_0)$.

4.2. *Derivation of constraint equations: The momentum constraint*

This subsection is a continuation of the previous subsection where we showed in detail the derivation of the Hamiltonian constraint. In this subsection, we derive the so-called momentum constraint equation (4.2).

First, let us recall the Codazzi equation (1.13) which is given by the following identity

$$\overline{\mathrm{Rm}}^{\perp}(X, Y)Z = (\overline{\nabla}_X \mathrm{II})(Y, Z) - (\overline{\nabla}_Y \mathrm{II})(X, Z).$$

Using this, we obtain after applying to $(X, Y, Z) = (e_i, Y, e_i)$ and summing over i from 1 to n the following

$$\sum_{i=1}^{n} \overline{\mathrm{Rm}}^{\perp}(e_i, Y)e_i = \sum_{i=1}^{n} (\overline{\nabla}_{e_i} \mathrm{II})(Y, e_i) - \sum_{i=1}^{n} (\overline{\nabla}_Y \mathrm{II})(e_i, e_i) \qquad (4.8)$$

for any tangent vector Y. First, we estimate the left hand side of (4.8). Since $\overline{\mathrm{Rm}}(e_i, Y)e_i$, for all $i \geqslant 1$, belong to the tangent space, there hold

$$\langle \overline{\mathrm{Rm}}(e_i, Y)e_i, e_0 \rangle = 0 \quad \forall i = \overline{1, n}.$$

Furthermore, by splitting $\overline{\mathrm{Rm}}(e_i, Y)e_i = \overline{\mathrm{Rm}}''(e_i, Y)e_i + \overline{\mathrm{Rm}}^{\perp}(e_i, Y)e_i$ and then taking the inner product with $e_0 = \vec{n}$, we can write

$$\left\langle \overline{\mathrm{Rm}}^{\perp}(e_i, Y)e_i, e_0 \right\rangle = \langle \overline{\mathrm{Rm}}(e_i, Y)e_i, e_0 \rangle \quad \forall i = \overline{1, n}. \qquad (4.9)$$

Moreover, by the skew symmetry property of the Riemann curvature tensor, i.e., $\overline{\mathrm{Rm}}_{ijkl} = -\overline{\mathrm{Rm}}_{ijlk}$, we know that

$$\langle \overline{\mathrm{Rm}}(e_0, e_i)e_0, e_0 \rangle = 0 \quad \forall i = \overline{1, n}$$

which immediately implies

$$\langle \overline{\mathrm{Rm}}(e_0, Y)e_0, e_0 \rangle = 0.$$

Hence, by the definition of $\mathrm{Ric}_{\bar{g}}(Y, e_0)$, we obtain

$$\mathrm{Ric}_{\bar{g}}(Y, e_0) = -\left\langle \overline{\mathrm{Rm}}(e_0, Y)e_0, e_0 \right\rangle + \sum_{i=1}^{n} \left\langle \overline{\mathrm{Rm}}(e_i, Y)e_0, e_i \right\rangle$$

$$= -\sum_{i=1}^{n} \left\langle \overline{\mathrm{Rm}}(e_i, Y)e_i, e_0 \right\rangle.$$

This and the relation (4.9) help us to conclude

$$\mathrm{Ric}_{\bar{g}}(Y, e_0) = -\sum_{i=1}^{n} \left\langle \overline{\mathrm{Rm}}^{\perp}(e_i, Y)e_i, e_0 \right\rangle.$$

Thanks to (4.8), we arrive at

$$\mathrm{Ric}_{\bar{g}}(Y, e_0) = -\sum_{i=1}^{n} \left\langle (\overline{\nabla}_{e_i} \mathrm{I\!I})(Y, e_i), e_0 \right\rangle + \sum_{i=1}^{n} \left\langle (\overline{\nabla}_Y \mathrm{I\!I})(e_i, e_i), e_0 \right\rangle. \quad (4.10)$$

We now evaluate the right hand side of (4.10). In order to achieve that goal, we make use of the formula involving the covariant derivative of the second fundamental form. Indeed, for the first term involving $\left\langle (\overline{\nabla}_{e_i} \mathrm{I\!I})(Y, e_i), e_0 \right\rangle$, we first write

$$(\overline{\nabla}_{e_i} \mathrm{I\!I})(Y, e_i) = \overline{\nabla}_{e_i}^{\perp}(\mathrm{I\!I}(Y, e_i)) - \mathrm{I\!I}(\nabla_{e_i} Y, e_i) - \mathrm{I\!I}(Y, \nabla_{e_i} e_i)$$

$$= -\overline{\nabla}_{e_i}^{\perp}(K(Y, e_i)e_0) + K(\nabla_{e_i} Y, e_i)e_0 + K(Y, \nabla_{e_i} e_i)e_0.$$

Using the Leibniz rule and thanks to $\left\langle \overline{\nabla}_{e_i} e_0, e_0 \right\rangle = 0$, we can write

$$\overline{\nabla}_{e_i}^{\perp}(K(Y, e_i)e_0) = \overline{\nabla}_{e_i}^{\perp}(K(Y, e_i))e_0 + K(Y, e_i)\underbrace{\overline{\nabla}_{e_i}^{\perp} e_0}_{=0}$$

$$= \overline{\nabla}_{e_i}(K(Y, e_i))e_0.$$

Thus, we get

$$\left\langle (\overline{\nabla}_{e_i} \mathrm{I\!I})(Y, e_i), e_0 \right\rangle = \overline{\nabla}_{e_i}(K(Y, e_i)) - K(\nabla_{e_i} Y, e_i) - K(Y, \nabla_{e_i} e_i).$$

By summing, we obtain

$$\sum_{i=1}^{n} \left\langle (\overline{\nabla}_{e_i} \mathrm{I\!I})(Y, e_i), e_0 \right\rangle = (\mathrm{div}_g K)(Y). \quad (4.11)$$

For the second term involving $\left\langle (\overline{\nabla}_Y \mathrm{I\!I})(e_i, e_i), e_0 \right\rangle$, we do the same way as follows

$$(\overline{\nabla}_Y \mathrm{I\!I})(e_i, e_i) = \overline{\nabla}_Y^{\perp}(\mathrm{I\!I}(e_i, e_i)) - \mathrm{I\!I}(\nabla_Y e_i, e_i) - \mathrm{I\!I}(e_i, \nabla_Y e_i)$$

$$= \overline{\nabla}_Y^{\perp}(\mathrm{I\!I}(e_i, e_i)) - 2\mathrm{I\!I}(\nabla_Y e_i, e_i).$$

Without loss of generality, we can assume that we are in normal coordinates at some point, say $p \in M$. This immediately implies that $\nabla_Y e_i = 0$ for all $i = \overline{1, n}$. Therefore,

$$
\begin{aligned}
(\overline{\nabla}_Y \mathbb{II})(e_i, e_i) &= \overline{\nabla}_Y^{\perp}(\mathbb{II}(e_i, e_i)) \\
&= -\overline{\nabla}_Y^{\perp}(K(e_i, e_i)e_0) \\
&= -\overline{\nabla}_Y^{\perp}(K(e_i, e_i))e_0 + K(e_i, e_i) \underbrace{\nabla_Y^{\perp} e_0}_{=0} \\
&= -\overline{\nabla}_Y(K(e_i, e_i))e_0.
\end{aligned}
$$

Thus, by summing, we obtain

$$
\sum_{i=1}^{n} \langle (\overline{\nabla}_Y \mathbb{II})(e_i, e_i), e_0 \rangle = \overline{\nabla}_Y \left(\sum_{i=1}^{n} K(e_i, e_i) \right) = d(\operatorname{tr}_g K)(Y). \qquad (4.12)
$$

Simply combining (4.10), (4.11), and (4.12), we have just proved that

$$
-\operatorname{Ric}_{\overline{g}}(Y, e_0) = (\operatorname{div}_g K)(Y) - d(\operatorname{tr}_g K)(Y).
$$

By using the Einstein equation and thanks to $\overline{g}(Y, e_0) = 0$, we get that

$$
\operatorname{Ric}_{\overline{g}}(Y, e_0) = \operatorname{Ric}_{\overline{g}}(Y, e_0) - \frac{1}{2} \underbrace{\overline{g}(Y, e_0)}_{=0} \operatorname{Scal}_{\overline{g}} = T(Y, e_0).
$$

Thus, we arrive at

$$
\operatorname{div}_g K - d(\operatorname{tr}_g K) = -T(\cdot, e_0).
$$

We note that, throughout the proof, we have used the fact that the vector $\nabla_Y e_0$ belongs to the tangent space, thus giving us $\nabla_Y^{\perp} e_0 = 0$.

4.3. Evolutions of the first and second fundamental forms

One of the interesting features of the Einstein equation is that it can be considered as a dynamical system. To understand this, we simply go back to (1.7). Assume that we have already had a spacetime $(\mathcal{M}, \overline{g}, \overline{\nabla})$ and that each slice $\Sigma_t = M \times \{t\}$ is spacelike in $\mathcal{M}$. Then whenever we know the induced metric g on Σ_t, the lapse function N, and the shift vector β, we can fully recover the spacetime metric $\overline{g}$. In other words, determining solutions of the initial value problems is equivalent to specifying the lapse and shift based on the initial data.

However, one can formulate the problem in the reverse way: Fixing a good choice of the lapse function and the shift vector together with an initial hypersurface M, we need to solve the first and second fundamental forms on slices using evolutions. The evolutions that we have just mentioned can be derived easily but lengthy computation cannot be avoided. To be exact, the induced metric g and the second fundamental form K obey the following system:

$$\frac{\partial}{\partial t} g_{ij} = -2NK_{ij} + \mathcal{L}_\beta g_{ij}$$

and

$$\frac{\partial}{\partial t} K_{ij} = -\nabla_i \nabla_j N + N(\overline{\text{Ric}}_{ij} - \text{Ric}_{ij} + KK_{ij} - 2K_{i\mu}K_j^\mu) + \mathcal{L}_\beta K_{ij}$$

where $\mathcal{L}_\beta$ is the Lie derivative along the vector field β.

4.4. *Construction of spacetimes via solutions of the vacuum Einstein constraint equations*

In this subsection, given a solution (g, K) of the constraint equations (4.1) and (4.2) on a manifold (M, g) of the dimension n, we shall construct an appropriate spacetime $(\mathcal{M}, \overline{g})$ of the dimension $n+1$ such that the Einstein equation (2.4) holds in the vacuum case.

4.4.1. *Construction of spacetimes*

First, we observe that $\text{tr}(\text{Ric}) = \text{Scal}$ and $\text{tr}(\overline{g}) = n + 1$. Hence, taking the trace both sides of (2.4) with respect to the spacetime metric $\overline{g}$ to get

$$\frac{1 - n}{2} \text{Scal}_{\overline{g}} = \text{tr}_{\overline{g}}(T).$$

Thus, we can rewrite (2.4) as follows

$$\text{Ric}_{\overline{g}} = T - \frac{\overline{g}}{n - 1} \text{tr}_{\overline{g}}(T). \tag{4.13}$$

Recall from (2.11) and using (4.13) we obtain the following identity

$$\overline{\text{Ric}}_{ij} = \frac{1}{2}(\overline{g}_{\alpha i}\partial_j \lambda^\alpha + \overline{g}_{\alpha j}\partial_i \lambda^\alpha) - \frac{1}{2}\overline{g}^{km}\overline{g}_{ij,km} + \text{lower orders}$$

$$= T_{ij} - \frac{1}{n - 1} \text{tr}_{\overline{g}}(T)\overline{g}_{ij}.$$

Note that in the vacuum case, we simply ignore all T. At the very beginning, the metric $\bar{g}$ can be fully determined by setting

$$\begin{cases} \bar{g}_{ij} = g_{ij}, & 1 \leqslant i, j \leqslant n, \\ \bar{g}_{00} = -1, \\ \bar{g}_{0j} = 0, & 1 \leqslant j \leqslant n, \\ \bar{g}_{ij,0} = -2K_{ij}, & 1 \leqslant i, j \leqslant n. \end{cases} \tag{4.14}$$

How about $\bar{g}_{ij,k}$? Clearly, as a consequence of the choice above, we also have $\bar{g}_{ij,k}$ for non-zero i, j, k. To fully construct the initial condition, we still need to find $\bar{g}_{0j,0}$ as they cannot be obtained from (4.14).

First, in view of the gauge condition, $\lambda^\alpha = \Box_{\bar{g}} x^\alpha = 0$, we find that

$$\lambda^\alpha = \frac{1}{\sqrt{|\det \bar{g}|}} \frac{\partial}{\partial x^j} \left(\sqrt{|\det \bar{g}|}\, \bar{g}^{ij} \frac{\partial}{\partial x^i} (x^\alpha) \right)$$

$$= \bar{g}^{\alpha j}_{,j} + \frac{1}{2} \bar{g}^{\alpha j} \bar{g}^{pq} \bar{g}_{pq,j} = 0.$$

Therefore, at $t = 0$, we get that

$$\begin{cases} \lambda^0 = \bar{g}^{00}_{,0} + \frac{1}{2} \bar{g}^{00} \bar{g}^{pq} \bar{g}_{pq,0}, \\ \lambda^\alpha = \bar{g}^{\alpha 0}_{,0} + \underbrace{\sum_{j=1}^{n} \left(\bar{g}^{\alpha j}_{,j} + \frac{1}{2} \bar{g}^{\alpha j} \bar{g}^{pq} \bar{g}_{pq,j} \right)}_{\text{already determined}}, & 1 \leqslant \alpha \leqslant n. \end{cases}$$

Hence, in order to guarantee that $\lambda^\alpha \equiv 0$ at $t = 0$ for all $\alpha = \overline{0, n}$, we first select $\bar{g}^{\alpha 0}_{,0}$ such that $\lambda^\alpha = 0$ at $t = 0$. Once this task is done, we can determine $\bar{g}^{00}_{,0}$ at $t = 0$ such that $\lambda^0 = 0$ at $t = 0$. In other words, the initial data that preserves the gauge condition can be found in this way.

Within a small time, the hyperbolic system mentioned above always admits a solution $\bar{g}$. However, it is not necessary to have that $\bar{g}$ solves the Einstein equations unless the gauge condition remains valid within the small time. We shall prove this affirmatively.

4.4.2. *The propagate of the gauge condition in time*

We now consider how could the gauge condition propagate in time so long as the metric $\bar{g}$ solves the reduced equations. As we shall see, all λ^α satisfy a homogeneous linear wave equation which is a consequence of the Bianchi identities with vanishing initial time derivatives which come from the constraint equations.

First, recall that in the vacuum setting, the Einstein equation is nothing but $\mathrm{Ric}_{\overline{g}} = 0$. We now introduce the following reduced Einstein equation

$$\overline{\mathrm{Ric}}_{ij} + \frac{1}{2}(\overline{g}_{\alpha i}\partial_j \lambda^\alpha + \overline{g}_{\alpha j}\partial_i \lambda^\alpha) = 0. \tag{4.15}$$

Clearly, whenever $\overline{g}$ solves (4.15) along with the gauge condition $\lambda^\alpha = 0$, we immediately conclude that $\overline{g}$ solves the original Einstein equation. Keep in mind that the term $\overline{g}_{\alpha i}\partial_j \lambda^\alpha + \overline{g}_{\alpha j}\partial_i \lambda^\alpha$ only vanishes at the initial time $t = 0$.

For simplicity, let us denote by L the following 2-tensor

$$L_{\alpha\beta} = \frac{1}{2}\overline{g}_{\beta\mu}\partial_\alpha \lambda^\mu + \frac{1}{2}\overline{g}_{\alpha\mu}\partial_\beta \lambda^\mu,$$

hence we can rewrite (4.15) as $\overline{\mathrm{Ric}}_{ij} = -L_{ij}$. Using this, we obtain

$$\overline{\mathrm{Ric}}^{\alpha\beta} - \frac{1}{2}\overline{g}^{\alpha\beta}\mathrm{Scal}_{\overline{g}} = -(L^{\alpha\beta} - \frac{1}{2}\overline{g}^{\alpha\beta}L),$$

where L is the trace of $L^{\alpha\beta}$ given by $L = \overline{g}_{ij}L^{ij}$. Since the Einstein tensor is divergence free, we know that $L^{\alpha\beta} - \frac{1}{2}\overline{g}^{\alpha\beta}L$ is also divergence-free; hence we obtain

$$\nabla_\gamma\left(L^{\gamma\mu} - \frac{1}{2}g^{\gamma\mu}L\right) = 0.$$

Keep in mind that $\nabla_\gamma \overline{g} = 0$. A simple calculation shows that

$$
\begin{aligned}
0 &= \nabla_\gamma(\overline{g}^{\alpha\gamma}\overline{g}^{\beta\mu}L_{\alpha\beta} - \frac{1}{2}\overline{g}^{\gamma\mu}\overline{g}^{\alpha\beta}L_{\alpha\beta}) \\
&= \frac{1}{2}\nabla_\gamma\left(\overline{g}^{\alpha\gamma}\overline{g}^{\beta\mu}(\overline{g}_{\beta\mu}\partial_\alpha\lambda^\mu + \overline{g}_{\alpha\mu}\partial_\beta\lambda^\mu) - \frac{1}{2}\overline{g}^{\gamma\mu}\overline{g}^{\alpha\beta}(\overline{g}_{\beta\mu}\partial_\alpha\lambda^\mu + \overline{g}_{\alpha\mu}\partial_\beta\lambda^\mu)\right) \\
&= \frac{1}{2}\overline{g}^{\alpha\gamma}\underbrace{\overline{g}^{\beta\mu}\overline{g}_{\beta\mu}}_{1}\nabla_\gamma(\partial_\alpha\lambda^\mu) + \frac{1}{2}\overline{g}^{\beta\mu}\underbrace{\overline{g}^{\alpha\gamma}\overline{g}_{\alpha\mu}}_{\delta^\gamma_\mu}\nabla_\gamma(\partial_\beta\lambda^\mu) \\
&\quad - \frac{1}{4}\overline{g}^{\gamma\mu}\underbrace{\overline{g}^{\alpha\beta}\overline{g}_{\beta\mu}}_{\delta^\alpha_\mu}\nabla_\gamma(\partial_\alpha\lambda^\mu) - \frac{1}{4}\overline{g}^{\gamma\mu}\underbrace{\overline{g}^{\alpha\beta}\overline{g}_{\alpha\mu}}_{\delta^\beta_\mu}\nabla_\gamma(\partial_\beta\lambda^\mu) \\
&= \frac{1}{2}\overline{g}^{\alpha\gamma}\nabla_\gamma(\partial_\alpha\lambda^\mu) + \frac{1}{2}\overline{g}^{\beta\mu}\nabla_\gamma(\partial_\beta\lambda^\gamma) - \frac{1}{2}\overline{g}^{\gamma\mu}\nabla_\gamma(\partial_\beta\lambda^\beta).
\end{aligned}
$$

Thanks to $\Box_{\overline{g}} = \overline{g}^{\gamma\alpha}\nabla_\gamma(\partial_\alpha(\cdot))$, we have proved that λ^α solves the following homogeneous linear hyperbolic system

$$\Box_{\overline{g}}\lambda^\mu = -\overline{g}^{\beta\mu}\nabla_\gamma(\partial_\beta\lambda^\gamma) + \overline{g}^{\gamma\mu}\nabla_\gamma(\partial_\beta\lambda^\beta).$$

The above long calculation also shows that

$$
\begin{aligned}
0 =&\ \overline{\mathrm{Ric}}_{\alpha\beta} - \frac{1}{2}\bar{g}_{\alpha\beta}\,\mathrm{Scal}_{\bar{g}}\\
=&\ -L_{\alpha\beta} + \frac{1}{2}\bar{g}_{\alpha\beta}L\\
=&\ -\frac{1}{2}\bar{g}_{\beta\mu}\partial_\alpha\lambda^\mu - \frac{1}{2}\bar{g}_{\alpha\mu}\partial_\beta\lambda^\mu\\
&\ +\frac{1}{2}\bar{g}_{\alpha\beta}\bar{g}^{pq}\left(\frac{1}{2}\bar{g}_{q\mu}\partial_p\lambda^\mu + \frac{1}{2}\bar{g}_{p\mu}\partial_q\lambda^\mu\right)\\
=&\ -\frac{1}{2}\bar{g}_{\beta\mu}\partial_\alpha\lambda^\mu - \frac{1}{2}\bar{g}_{\alpha\mu}\partial_\beta\lambda^\mu\\
&\ +\frac{1}{4}\bar{g}_{\alpha\beta}\underbrace{\bar{g}^{pq}\bar{g}_{q\mu}}_{\delta_\mu^p}\partial_p\lambda^\mu + \frac{1}{4}\bar{g}_{\alpha\beta}\underbrace{\bar{g}^{pq}\bar{g}_{p\mu}}_{\delta_\mu^q}\partial_q\lambda^\mu\\
=&\ -\frac{1}{2}\bar{g}_{\beta\mu}\partial_\alpha\lambda^\mu - \frac{1}{2}\bar{g}_{\alpha\mu}\partial_\beta\lambda^\mu + \frac{1}{2}\bar{g}_{\alpha\beta}\partial_\mu\lambda^\mu.
\end{aligned}
$$

Since we are in the vacuum case, at $t = 0$, we get

$$
-\frac{1}{2}\bar{g}_{0\mu}\partial_\alpha\lambda^\mu - \frac{1}{2}\bar{g}_{\alpha\mu}\partial_0\lambda^\mu + \frac{1}{2}\bar{g}_{\alpha 0}\partial_\mu\lambda^\mu = 0.
$$

By using $\mu = \alpha = 0$, i.e. the Hamiltonian constraint (4.1), we obtain $\partial_0\lambda^0 = 0$. We still need to prove that $\partial_0\lambda^\mu = 0$ for all $\mu = \overline{1,n}$. Indeed, thanks to $\partial_\alpha\lambda^\mu = 0$ for any $\alpha, \mu > 0$, we know that

$$
\begin{aligned}
0 =&\ -\frac{1}{2}\bar{g}_{0\mu}\underbrace{\partial_\alpha\lambda^\mu}_{0} - \frac{1}{2}\bar{g}_{\alpha\mu}\partial_0\lambda^\mu + \frac{1}{2}\bar{g}_{\alpha 0}\underbrace{\partial_\mu\lambda^\mu}_{0}\\
=&\ -\frac{1}{2}\sum_{\mu=1}^{n}\bar{g}_{\alpha\mu}\partial_0\lambda^\mu.
\end{aligned}
$$

Since this is true for all $\alpha = \overline{1,n}$ and the matrix $(\bar{g}_{\alpha\mu})$ is invertible, we have that all $\partial_0\lambda^\alpha$ vanish at $t = 0$ for all $\alpha > 0$.

Since λ^α satisfy a homogeneous linear hyperbolic system with vanishing initial data, λ^α vanish identically by the uniqueness for solutions of the Cauchy problem for the hyperbolic evolutions. This proves how the gauge condition $\lambda^\alpha = 0$ is preserved in (small) time.

5. Conformal Method and Transformed PDEs

In the literature, there are two routes that have been followed to solve the system of constraints. First, based on perturbation methods, people perform a connected sum between known solutions of the constraints in order to produce new solutions which obey some properties. Among others, we can list a few works such as [Cor00, CS06] by Corvino and Schoen, [CD03] by Chruściel and Delay.

The other route introduced by Lichnerowicz, Choquet-Bruhat, and York is the so-called conformal method. In this section, we show how to use the conformal method to derive the recast constraint equations in the dimension $n \geqslant 2$. It is interesting to know that the causal structure of spacetimes only depend on their Lorentzian metric up to a conformal factor; hence, when dealing with conformal metrics, it suffices to play with conformal factor.

First, we assume that the given initial data $(M, \tilde{g}, \tilde{K})$ fulfills the constraint equations (4.1) and (4.2). In this new notation, we rewrite (4.1) and (4.2) as the following

$$\mathrm{Scal}_{\tilde{g}} - |\tilde{K}|_{\tilde{g}}^2 + (\mathrm{tr}_{\tilde{g}} \tilde{K})^2 = 2\tilde{\rho} \tag{5.1}$$

and

$$\mathrm{div}_{\tilde{g}} \tilde{K} - d(\mathrm{tr}_{\tilde{g}} \tilde{K}) = \tilde{J}, \tag{5.2}$$

where $\tilde{\rho} = T(e_0, e_0)$ and $\tilde{J} = -T(\cdot, e_0)$. Note that in this context, we again denote $\tau = \mathrm{tr}_{\tilde{g}} \tilde{K}$.

5.1. *Conformal method*

As can be seen, the couple of constraint equations (5.1) and (5.2) for the initial data $(M, \tilde{g}, \tilde{K})$ consist of $n + 1$ equations (a scalar equation and an n-vector equation) with more than $n+1$ unknowns (for example, the metric $\tilde{g}$ consists of $n(n + 1)/2$ components). Since the constraint equations form an under-determined system, they are in general hard to solve. Fortunately, we have a technique known as the conformal method that can be used in most of cases, see [CBY80].

Definition 5.1: Two metrics g and $\tilde{g}$ on the manifold M of dimension $n \geqslant 2$ are called conformal if there exists some smooth function φ such that $\tilde{g} = e^{2\varphi} g$.

In local coordinates, the condition $\tilde{g} = e^{2\varphi} g$ is nothing but $\tilde{g}_{ij} = e^{2\varphi} g_{ij}$. Therefore, if we denote by $(\tilde{g}^{ij})$ and (g^{ij}) the inverse metric of $(\tilde{g}_{ij})$ and (g_{ij}) respectively, then we obtain $\tilde{g}^{ij} = e^{-2\varphi} g^{ij}$. Note that elementary calculations show $\tilde{g}^{ij} \tilde{g}_{ij} = g^{ij} g_{ij} = n$.

The basic idea of the conformal method is to equalize the number of equations and the number of unknowns in such a way that the resulting system is determined. More specific, the idea of the conformal method is to split the set of initial data $(M, \widetilde{g}, \widetilde{K})$ into the following two catalogues:

- *Conformal data*: Degrees of freedom that can be freely chosen; and
- *Determined data*: Degrees of freedom that are to be found by solving a determined system of partial differential equations.

We now discuss the method more precise. But first, it is well-known that the scalar curvatures $\mathrm{Scal}_{\widetilde{g}}$ and Scal_g of the conformal metrics $\widetilde{g}$ and g are related by the following rule

$$\mathrm{Scal}_{\widetilde{g}} = e^{-2\varphi}\big(\mathrm{Scal}_g - 2(n-1)\Delta_g\varphi - (n-1)(n-2)g^{ij}\partial_i\varphi\partial_j\varphi\big). \quad (5.3)$$

5.2. Transformed PDEs in the case $n \geqslant 3$

5.2.1. The transformed Hamiltonian constraint

When $n \geqslant 3$, we set $e^{2\varphi} = u^{2p}$ for some $u > 0$ and we choose p in such a way that the operator on u appearing within the brackets is somewhat linear in u; this goal is attained by choosing $p = \frac{2}{n-2}$, that is, $\widetilde{g} = u^{\frac{4}{n-2}}g$ which we suppose from now on. To see this, one immediately has $u = e^{\varphi/p}$ which implies after a direct computation that

$$\Delta_g u = \frac{1}{p^2}g^{ij}u\partial_i\varphi\partial_j\varphi + \frac{1}{p}u\Delta_g\varphi.$$

Thus,

$$g^{ij}\partial_i\varphi\partial_j\varphi = u^{-1}p^2\Delta_g u - p\Delta_g\varphi.$$

Hence,

$$\begin{aligned}
-2(n-1)\Delta_g\varphi &- (n-1)(n-2)g^{ij}\partial_i\varphi\partial_j\varphi \\
&= -(n-1)(2 - p(n-2))\Delta_g\varphi - (n-1)(n-2)u^{-1}p^2\Delta_g u \\
&= -\frac{4(n-1)}{n-2}u^{-1}\Delta_g u
\end{aligned}$$

provided $p = \frac{2}{n-2}$. Thus, Eq. (5.3) becomes

$$\mathrm{Scal}_{\widetilde{g}} = u^{-\frac{n+2}{n-2}}\Big(u\,\mathrm{Scal}_g - \frac{4(n-1)}{n-2}\Delta_g u\Big).$$

The Hamiltonian constraint (4.1) now becomes a semilinear elliptic equation for u given below

$$\frac{4(n-1)}{n-2}\Delta_g u - \mathrm{Scal}_g u + (|\widetilde{K}|_{\widetilde{g}}^2 - \tau^2 + 2\widetilde{\rho})u^{\frac{n+2}{n-2}} = 0. \quad (5.4)$$

As can be observed from (5.4), we are left with the term $|\widetilde{K}|_{\widetilde{g}}$. To transform this term, we shall use the transverse-traceless decomposition. However, we first need to study the momentum constraint.

5.2.2. *The transformed momentum constraint*

We still use the fact that $\widetilde{g} = u^{\frac{4}{n-2}} g$. If we denote by $\text{div}_{\widetilde{g}}$ and div_g the divergences in the metrics $\widetilde{g}$ and g respectively, we then have the following useful result, see [CBru09, Lemma 3.1].

Lemma 5.2: *On an n-dimensional manifold, if $\widetilde{g} = u^{\frac{4}{n-2}} g$, the covariant derivatives in $\widetilde{g}$ and g being respectively denoted $\widetilde{\nabla}$ and ∇, the divergences in the metrics $\widetilde{g}$ and g of an arbitrary contravariant 2-tensor P^{ij} verify the following identity*

$$\text{div}_{\widetilde{g}}\, P^{ij} = u^{-\frac{2(n+2)}{n-2}}\, \text{div}_g(u^{\frac{2(n+2)}{n-2}}\, P^{ij}) - \frac{2}{n-2}\frac{\partial_i u}{u} g^{ij}\, \text{tr}_g P. \qquad (5.5)$$

Proof: For the purpose of clarity we may denote the tensor P by

$$P = P^{ij} \frac{\partial}{\partial x^i} \otimes \frac{\partial}{\partial x^j}.$$

Using the Leibniz rule, one easily gets

$$(\widetilde{\nabla} P)\left(\cdot, \cdot, \frac{\partial}{\partial x^k}\right) = \widetilde{\nabla}_{\frac{\partial}{\partial x^k}}\left(P^{ij} \frac{\partial}{\partial x^i} \otimes \frac{\partial}{\partial x^j}\right)$$

$$= \frac{\partial P^{ij}}{\partial x^k} \frac{\partial}{\partial x^i} \otimes \frac{\partial}{\partial x^j} + P^{ij}\left(\widetilde{\nabla}_{\frac{\partial}{\partial x^k}} \frac{\partial}{\partial x^i}\right) \otimes \frac{\partial}{\partial x^j}$$

$$+ P^{ij} \frac{\partial}{\partial x^i} \otimes \left(\widetilde{\nabla}_{\frac{\partial}{\partial x^k}} \frac{\partial}{\partial x^j}\right)$$

$$= \left(\frac{\partial P^{ij}}{\partial x^k} + P^{lj}\widetilde{\Gamma}^i_{lk} + P^{il}\widetilde{\Gamma}^j_{lk}\right) \frac{\partial}{\partial x^i} \otimes \frac{\partial}{\partial x^j}.$$

Therefore, $(2, 1)$-tensor $\widetilde{\nabla} P$, which is of the form

$$\widetilde{\nabla} P = \widetilde{\nabla}_k P^{ij} \frac{\partial}{\partial x^i} \otimes \frac{\partial}{\partial x^j} \otimes dx^k = P^{ij}_{;k} \frac{\partial}{\partial x^i} \otimes \frac{\partial}{\partial x^j} \otimes dx^k,$$

verifies

$$\widetilde{\nabla}_k P^{ij} = P^{ij}_{;k} = \partial_k P^{ij} + P^{lj}\widetilde{\Gamma}^i_{lk} + P^{il}\widetilde{\Gamma}^j_{lk}.$$

We take the divergence, that is, to use

$$\text{div}\, P = \delta^k_i P^{ij}_{,k} \frac{\partial}{\partial x^j} = \delta^k_i \widetilde{\nabla}_k P^{ij} \frac{\partial}{\partial x^j} = \widetilde{\nabla}_i P^{ij} \frac{\partial}{\partial x^j}$$

to arrive at

$$\widetilde{\nabla}_i P^{ij} = \partial_i P^{ij} + P^{lj}\widetilde{\Gamma}^i_{li} + P^{il}\widetilde{\Gamma}^j_{li}. \tag{5.6}$$

Notice that under the conformal change $\widetilde{g}_{ij} = u^{\frac{4}{n-2}} g_{ij}$, the Christoffel symbols computed with respect to $\widetilde{g}$ and g verify the following identity

$$\widetilde{\Gamma}^k_{ij} = \Gamma^k_{ij} + \frac{2}{n-2}\frac{1}{u}(\delta^k_j \partial_i u + \delta^k_i \partial_j u - g^{kl} g_{ij} \partial_l u).$$

Therefore,

$$\begin{aligned}
\widetilde{\nabla}_i P^{ij} &= \partial_i P^{ij} + P^{lj}\Big[\Gamma^i_{li} + \frac{2}{n-2}\frac{1}{u}(\delta^i_l \partial_i u + \delta^i_i \partial_l u - g^{ik} g_{li} \partial_k u)\Big] \\
&\quad + P^{il}\Big[\Gamma^j_{li} + \frac{2}{n-2}\frac{1}{u}(\delta^j_l \partial_i u + \delta^j_i \partial_l u - g^{jk} g_{li} \partial_k u)\Big] \\
&= \nabla_i P^{ij} + \frac{2}{n-2}\frac{1}{u}(P^{lj}\partial_l u + n P^{lj}\partial_l u - P^{lj}\partial_l u) \\
&\quad + \frac{2}{n-2}\frac{1}{u}(P^{lj}\partial_l u + P^{jl}\partial_l u - P^{il} g^{jk} g_{li} \partial_k u) \\
&= u^{-\frac{2(n+2)}{n-2}} \nabla_i\big(u^{\frac{2(n+2)}{n-2}} P^{ij}\big) - \frac{2}{n-2}\frac{\partial_k u}{u} g^{jk}(g_{li} P^{il}).
\end{aligned}$$

The proof immediately follows thanks to $\text{tr}_g P = g_{li} P^{il}$.   □

In view of (5.5), it is convenient to split the unknown $\widetilde{K}$ into a weighted traceless part and its trace with respect to the conformal metric g, namely, we write

$$\widetilde{K}^{ij} = u^{-\frac{2(n+2)}{n-2}} \widehat{K}^{ij} + \frac{\tau}{n}\widetilde{g}^{ij}.$$

By lowering the indices in $\overline{K}$ and $\widetilde{K}$, one gets

$$\widetilde{K}_{ij} = u^{-2}\widehat{K}_{ij} + \frac{\tau}{n}\widetilde{g}_{ij}.$$

It is clear to see that the tensor $\widehat{K}$ is symmetric and traceless with respect to g, that is

$$\text{tr}_g \widehat{K} = g^{ij}\widehat{K}_{ij} = u^{-\frac{4}{n-2}}\widetilde{g}^{ij}\widehat{K}_{ij} = u^{\frac{2n}{n-2}}\widetilde{g}^{ij}\Big(\widetilde{K}_{ij} - \frac{\tau}{n}\widetilde{g}_{ij}\Big) = 0,$$

where we have used $\text{tr}_{\widetilde{g}}\widetilde{K} = \tau$ and $\widetilde{g}^{ij}\widetilde{g}_{ij} = n$. In view of the momentum constraint (5.2) and with the fact that

$$u^{-\frac{4}{n-2}}\tau = u^{-\frac{4}{n-2}} \text{tr}_{\widetilde{g}}\widetilde{K} = \text{tr}_g \widetilde{K},$$

and that $\operatorname{div}_{\widetilde{g}} \widetilde{K} = \widetilde{\nabla}_i \widetilde{K}^i$, we have by (5.5)

$$\widetilde{J}^j + \widetilde{g}^{ij} \partial_i \tau = \widetilde{\nabla}_i \widetilde{K}^{ij}$$

$$= u^{-\frac{2(n+2)}{n-2}} \nabla_i (u^{\frac{2(n+2)}{n-2}} \widetilde{K}^{ij}) - \frac{2}{n-2} \frac{\partial_i u}{u} g^{ij} \operatorname{tr}_g \widetilde{K}$$

$$= u^{-\frac{2(n+2)}{n-2}} \nabla_i (\widehat{K}^{ij} + \frac{1}{n} u^{\frac{2(n+2)}{n-2}} \widetilde{g}^{ij} \tau) - \frac{2}{n-2} \frac{\partial_i u}{u} \widetilde{g}^{ij} \operatorname{tr}_{\widetilde{g}} \widetilde{K}$$

$$= u^{-\frac{2(n+2)}{n-2}} \nabla_i \widehat{K}^{ij} + \frac{1}{n} u^{-\frac{2(n+2)}{n-2}} \nabla_i (u^{\frac{2n}{n-2}} g^{ij} \tau) - \frac{2}{n-2} \frac{\partial_i u}{u} \widetilde{g}^{ij} \tau$$

$$= u^{-\frac{2(n+2)}{n-2}} \nabla_i \widehat{K}^{ij} + \frac{1}{n} \underbrace{g^{ij} u^{-\frac{2(n+2)}{n-2}} \nabla_i (u^{\frac{2n}{n-2}} \tau)}_{\widetilde{g}^{ij} u^{-\frac{2n}{n-2}}} - \frac{2}{n-2} \frac{\partial_i u}{u} \widetilde{g}^{ij} \tau$$

$$= u^{-\frac{2(n+2)}{n-2}} \nabla_i \widehat{K}^{ij} + \frac{1}{n} \widetilde{g}^{ij} \nabla_g \tau.$$

Thus, we have transformed $\widetilde{J}^j + \widetilde{g}^{ij} \partial_i \tau = \widetilde{\nabla}_i \widetilde{K}^{ij}$ to the following equation

$$\nabla_i \widehat{K}^{ij} = u^{\frac{2(n+2)}{n-2}} \widetilde{J}^j + \frac{n-1}{n} u^{\frac{2(n+2)}{n-2}} \widetilde{g}^{ij} \partial_i \tau.$$

Equivalently, thanks to $\widetilde{g}_{ij} = u^{\frac{4}{n-2}} g_{ij}$, we further obtain

$$\nabla_i \widehat{K}^{ij} = \frac{n-1}{n} u^{\frac{2n}{n-2}} g^{ij} \partial_i \tau + u^{\frac{2(n+2)}{n-2}} \widetilde{J}^j.$$

In other words, the momentum constraint (5.2) now becomes

$$\operatorname{div}_g \widehat{K} = \frac{n-1}{n} u^{\frac{2n}{n-2}} \nabla_g \tau + u^{\frac{2(n+2)}{n-2}} \widetilde{J}. \tag{5.7}$$

In particular, if $\widetilde{J}$ is zero and τ is constant, the symmetric 2-tensor $\widehat{K}$ is also divergence free. Symmetric 2-tensors which are divergence free and trace free are called TT-tensors (tranverse, traceless). Let us go back to the decomposition of $\widetilde{K}$. Clearly,

$$|\widetilde{K}|_{\widetilde{g}}^2 = \widetilde{g}_{ih} \widetilde{g}_{jk} \widetilde{K}^{ij} \widetilde{K}^{hk}$$

$$= u^{-\frac{4n}{n-2}} g_{ih} g_{jk} \widehat{K}^{ij} \widehat{K}^{hk} + \frac{\tau^2}{n}$$

$$= u^{-\frac{4n}{n-2}} |\widehat{K}|_g^2 + \frac{\tau^2}{n}.$$

Therefore, Eq. (5.4) now reads as follows

$$\frac{4(n-1)}{n-2} \Delta_g u - \operatorname{Scal}_g u + |\widehat{K}|_g^2 u^{\frac{-3n+2}{n-2}} - \left(\frac{n-1}{n} \tau^2 - 2\widetilde{\rho} \right) u^{\frac{n+2}{n-2}} = 0. \tag{5.8}$$

Clearly, this is a semilinear elliptic equation for u.

5.2.3. Scaling of the scalar fields

Recall that M is an n-dimensional manifold with spatial metric g. We denote by a tilde the values induced on M by the spacetime quantities. For the scalar field $\overline{\psi}$, there is no need to do any time and space decomposition. However, the wave equation that $\overline{\psi}$ fulfills, that is

$$\nabla^\alpha \nabla_\alpha \overline{\psi} = U'(\overline{\psi}),$$

suggests that the initial data for the scalar field $\overline{\psi}$ should be the induced function and normalized time derivative of the function on M. Based on this reason and for the sake of simplicity, we denote

$$\widetilde{\psi} = \overline{\psi}\big|_M$$

and we use $\widetilde{\pi}$ to denote the normalized time derivative of $\widetilde{\psi}$ restricted to M, that is,

$$\widetilde{\pi} = \frac{1}{N}\left(\frac{\partial}{\partial t}\widetilde{\psi} - \beta^i \frac{\partial}{\partial x^i}\widetilde{\psi}\right) = N^{-1}\widetilde{\partial_0 \psi}$$

where $\widetilde{\partial_0 \psi}$ is the value of $\partial_0 \psi$ on M. Due to the fact that the background metric g is unphysical, we associate it to an unphysical lapse $\widehat{N}$ so that N and $\widehat{N}$ have the same associated densitized lapse, that is

$$\frac{N}{\sqrt{\det \widetilde{g}}} = \frac{\widehat{N}}{\sqrt{\det g}}. \tag{5.9}$$

(Here by densitized lapse we mean the lapse function divided by the square root of the determinant of the spatial metric[d].) Thanks to the conformal change $\widetilde{g} = u^{\frac{4}{n-2}}g$, this condition is equivalent to

$$N = u^{\frac{2n}{n-2}}\widehat{N}.$$

In this setting, we now denote $\pi = \widehat{N}^{-1}\widetilde{\partial_0 \psi}$ for the initial data π. Therefore, π and $\widetilde{\pi}$ are related by the following scaling

$$\widetilde{\pi} = N^{-1}\widetilde{\partial_0 \psi} = u^{-\frac{2n}{n-2}}\pi.$$

---

[d]It is interesting to note that as a corollary of the Cramer rule, there holds

$$N\sqrt{|\det \widetilde{g}|} = \sqrt{|\det \overline{g}|}.$$

It is a good exercise to compare $\det \widetilde{g}$ and $\det g$ when $\widetilde{g} = u^{\frac{4}{n-2}}g$.

Next, in order to obtain precise formulas for $\widetilde{\rho}$ and $\widetilde{J}$, we recall that $\widetilde{\rho} = T(\vec{n}, \vec{n})$ and $\widetilde{J} = -T(\cdot, \vec{n})$; hence it suffices to obtain the spacetime metric $\bar{g}$. Clearly,

$$\bar{g}_{00} = \bar{g}(\partial_0, \partial_0) = N^2 \bar{g}(\vec{n}, \vec{n}) = -N^2$$

and

$$\bar{g}_{i0} = \bar{g}_{0i} = \bar{g}(\partial_0, \partial_i) = \bar{g}(N\vec{n}, \partial_i) = 0$$

since $\vec{n}$ and ∂_i are perpendicular. Hence, we can write

$$\bar{g} = (\bar{g}_{\alpha\beta}) = \begin{pmatrix} \bar{g}_{00} & \bar{g}_{01} & \cdots & \bar{g}_{0n} \\ \bar{g}_{10} & \bar{g}_{11} & \cdots & \vdots \\ \vdots & \vdots & \ddots & \vdots \\ \bar{g}_{n0} & \cdots & \cdots & \bar{g}_{nn} \end{pmatrix} = \begin{pmatrix} -N^2 & 0 & \cdots & 0 \\ 0 & g_{11} & \cdots & g_{1n} \\ \vdots & \vdots & \ddots & \vdots \\ 0 & g_{n1} & \cdots & g_{nn} \end{pmatrix}.$$

An elementary calculation shows that the inverse metric $\bar{g}^{\alpha\beta}$ is

$$(\bar{g}^{\alpha\beta}) = \begin{pmatrix} -\frac{1}{N^2} & 0 & \cdots & 0 \\ 0 & g^{11} & \cdots & g^{1n} \\ \vdots & \vdots & \ddots & \vdots \\ 0 & g^{n1} & \cdots & g^{nn} \end{pmatrix}.$$

Hence, recalling the following formula for the stress-energy tensor in (2.6)

$$T_{\alpha\beta} = \partial_\alpha \overline{\psi} \partial_\beta \overline{\psi} - \frac{1}{2} \bar{g}_{\alpha\beta} \partial_\mu \overline{\psi} \partial^\mu \overline{\psi} - \bar{g}_{\alpha\beta} U(\overline{\psi})$$

we easily obtain

$$\begin{aligned} T(n, n) &= N^{-2} T(\partial_0, \partial_0) \\ &= N^{-2} T_{00} \\ &= N^{-2} \left(\partial_0 \overline{\psi} \partial_0 \overline{\psi} + \frac{1}{2} N^2 \partial_\mu \overline{\psi} \partial^\mu \overline{\psi} - N^2 U(\overline{\psi}) \right) \\ &= N^{-2} \partial_0 \overline{\psi} \partial_0 \overline{\psi} + \frac{1}{2} \partial_\mu \overline{\psi} \partial^\mu \overline{\psi} - U(\overline{\psi}). \end{aligned}$$

In the same way, we also obtain

$$T(\partial_i, n) = N^{-1}T(\partial_i, \partial_0)$$
$$= N^{-1}(\partial_i\overline{\psi}\partial_0\overline{\psi} - \frac{1}{2}\overline{g}_{i0}\partial_\mu\overline{\psi}\partial^\mu\overline{\psi} - \overline{g}_{i0}U(\overline{\psi}))$$
$$= N^{-1}\partial_i\overline{\psi}\partial_0\overline{\psi}.$$

Thus, the energy density on M of a scalar field $\widetilde{\psi}$ with potential $U(\cdot)$ is

$$\widetilde{\rho} = \frac{1}{2}(|\widetilde{\pi}|^2 + \widetilde{g}^{ij}\partial_i\psi\partial_j\psi) + U(\psi)$$

where ψ and $\partial_i\psi$ are the values of $\widetilde{\psi}$ and $\partial_i\widetilde{\psi}$ on M, respectively. In terms of the initial data set, $\widetilde{\rho}$ becomes

$$\widetilde{\rho} = \frac{1}{2}(u^{-\frac{4n}{n-2}}|\pi|^2 + u^{-\frac{4}{n-2}}|\nabla\psi|_g^2) + U(\psi).$$

Hence, we can regroup (5.8) as follows

$$\frac{4(n-1)}{n-2}\Delta_g u - (\text{Scal}_g - |\nabla\psi|_g^2)u$$
$$+ (|\widehat{K}|_g^2 + |\pi|^2)u^{-\frac{3n-2}{n-2}} - \left(\frac{n-1}{n}\tau^2 - 2U(\psi)\right)u^{\frac{n+2}{n-2}} = 0. \quad (5.10)$$

Now for the momentum density, from the above computation, one can see that

$$\widetilde{J}^i = -N^{-1}\widetilde{g}^{ij}\partial_j\psi\,\widetilde{\partial_0\psi} = -u^{-\frac{2(n+2)}{n-2}}g^{ij}\partial_j\psi\pi.$$

That is equivalent to

$$\widetilde{J} = -u^{-\frac{2(n+2)}{n-2}}\pi\nabla_g\psi.$$

Using this formula for $\widetilde{J}$, Eq. (5.7) becomes

$$\text{div}_g\,\widehat{K} = \frac{n-1}{n}u^{\frac{2n}{n-2}}\nabla_g\tau - \pi\nabla_g\psi. \quad (5.11)$$

5.2.4. Conformal Killing operator and conformal vector Laplacian

As can be observed from Eqs. (5.10) and (5.11), we are left with studying terms involving $\widehat{K}$. This is possible by using the transverse-traceless decomposition in the next subsection. As a first step to study such a decomposition, we mention the so-called conformal Killing operator and its associated conformal vector Laplacian.

First, we start with its definition. Roughly speaking, the conformal killing operator is a generalization of the Killing operator relative to the metric g. It maps any vector field on M to some tensor of type $(2,0)$. More precisely, in components, we have

$$(\mathbb{L}_g W)^{ij} = \nabla^i W^j + \nabla^j W^i - \frac{2}{n} \nabla_k W^k g^{ij}, \qquad (5.12)$$

where $W = W^i \partial_i$ is a vector field on M. Immediately, one can check that $\mathbb{L}_g W$ is traceless as can be seen in the following

$$g_{ij}(\mathbb{L}_g W)^{ij} = g_{ij}\nabla^i W^j + g_{ij}\nabla^j W^i - 2\nabla_k W^k = 0.$$

We now define the so-called Conformal Vector Laplacian $\Delta_{g,\mathbb{L}}$ associated to the metric g. Formally, we define $\Delta_{g,\mathbb{L}} = \mathrm{div}_g\left(\mathbb{L}_g \cdot \right)$ which, in local coordinates, basically says that

$$(\Delta_{g,\mathbb{L}} W)^i = \nabla_j(\mathbb{L}_g W)^{ij}.$$

In components, we have

$$
\begin{aligned}
(\Delta_{g,\mathbb{L}} W)^i &= \nabla_j \nabla^i W^j + \nabla_j \nabla^j W^i - \frac{2}{n}\nabla^i \nabla_k W^k \\
&= \nabla^i \nabla_j W^j + \mathrm{Ric}^i_j\, W^j + \nabla_j \nabla^j W^i - \frac{2}{n}\nabla^i \nabla_k W^k \\
&= \frac{n-2}{n}\nabla^i \nabla_j W^j + \mathrm{Ric}^i_j\, W^j + \nabla_j \nabla^j W^i.
\end{aligned}
$$

Let us now determine the kernel of $\Delta_{g,\mathbb{L}}$. By definition, one can easily see that

$$\ker\left(\mathbb{L}_g\right) \subset \ker\left(\Delta_{g,\mathbb{L}}\right).$$

However, we shall prove that they are actually the same. For any vector field W on M, we first have

$$
\begin{aligned}
\int_M W_j(\Delta_{g,\mathbb{L}} W)^j &= \int_M W_j \nabla_l(\mathbb{L}_g W)^{jl} \\
&= \int_M \nabla_l\left[W_j(\mathbb{L}_g W)^{jl}\right] - (\mathbb{L}_g W)^{jl}\nabla_l W_j \\
&= -\int_M (\mathbb{L}_g W)^{jl}\nabla_l W_j,
\end{aligned}
$$

where the Gauss–Ostrogradsky theorem has been used to get the last line.

In view of the right hand side integrand, we see that

$$g_{ij}g_{kl}(\mathbb{L}_gW)^{ik}(\mathbb{L}_gW)^{jl}$$

$$= g_{ij}g_{kl}[\nabla^iW^k + \nabla^kW^i](\mathbb{L}_gW)^{jl} - \frac{2}{n}\nabla_mW^m \underbrace{g^{ik}g_{ij}g_{kl}}_{g_{ij}\delta^i_l}(\mathbb{L}_gW)^{jl}$$

$$= [g_{kl}\nabla_jW^k + g_{ij}\nabla_lW^i](\mathbb{L}_gW)^{jl} - \frac{2}{n}\nabla_mW^m \underbrace{g_{jl}(\mathbb{L}_gW)^{jl}}_{0}$$

$$= 2g_{ij}\nabla_lW^i(\mathbb{L}_gW)^{jl},$$

where we have used the symmetry and the traceless property of $(\mathbb{L}_gW)^{ij}$. Therefore, we can write

$$\int_M W_j(\Delta_{g,\text{conf}}W)^j = -\frac{1}{2}\int_M g_{ij}g_{kl}(\mathbb{L}_gW)^{ik}(\mathbb{L}_gW)^{jl}$$

$$= -\frac{1}{2}\int_M \|\mathbb{L}_gW\|_g^2.$$

Therefore, assume that $W \in \ker\left(\Delta_{g,\mathbb{L}}\right)$; hence $(\Delta_{g,\mathbb{L}}W)^i = 0$, we then see from the preceding equality that $\mathbb{L}_gW = 0$ since the metric g is positive definite. Hence, $W \in \ker(\mathbb{L}_g)$ as claimed.

For future benefit, we suppose that M has boundary ∂M. By integration by parts, we have the following

$$\int_M U_j(\Delta_{g,\mathbb{L}}W)^j = \int_M U_j\nabla_l(\mathbb{L}_gW)^{jl}$$

$$= \int_M \nabla_l[u_j(\mathbb{L}_gW)^{jl}] - (\mathbb{L}_gW)^{jl}\nabla_lU_j$$

$$= \int_{\partial M} (\mathbb{L}_gW)^{jl}\nu_jU_l - \int_M (\mathbb{L}_gW)^{jl}\nabla_lU_j,$$

for any vector fields $U = U^i\partial_i$ and W and ν is the outward normal vector field on ∂M. As above, we can write

$$\int_M U_j(\Delta_{g,\mathbb{L}}W)^j = \int_{\partial M} (\mathbb{L}_gW)^{jl}\nu_jU_l - \frac{1}{2}\int_M (\mathbb{L}_gW)^{jl}(\mathbb{L}_gU)_{jl}. \quad (5.13)$$

Let us now discuss the existence and uniqueness of solutions to the conformal vector Poisson equation

$$(\Delta_{g,\mathbb{L}}W)^i = S^i \quad (5.14)$$

where the vector field S is already given. We shall provide a necessary condition for solution of (5.14) to exist as shown below.

Proposition 5.3: *Assume that (M, g) is compact without boundary. Then a necessary condition for solution of (5.14) to exist is that the source S^i must be orthogonal to any vector field in the kernel $\ker(\Delta_{g,\mathbb{L}})$, in the sense that*

$$\int_M C_j S^j = 0 \quad \forall C \in \ker(\Delta_{g,\mathbb{L}}). \tag{5.15}$$

Proof: This is clear as the following. First, for any $C \in \ker(\Delta_{g,\mathbb{L}})$, we know from (5.13) that

$$\int_M C_j S^j = \int_M C_j (\Delta_{g,\mathbb{L}} W)^j$$
$$= -\frac{1}{2} \int_M g_{ij} g_{kl} (\mathbb{L}_g C)^{ik} (\mathbb{L}_g W)^{jl}.$$

Because $(\mathbb{L}_g C)^{ik} = 0$, we obtain the desired identity. □

Clearly if (M, g) admits no conformal Killing vector, the identity is trivial. To see why this condition also gives us a sufficient condition, we can use the Fredholm theory.

5.2.5. *The transverse-traceless decomposition*

We consider in this subsection the solvability of (5.11) using the transverse-traceless decomposition. Roughly speaking, we search for $\widehat{K}$ of the form

$$\widehat{K} = \widehat{K}_{TT} + \mathbb{L}_g W \tag{5.16}$$

where $\widehat{K}_{TT}$ is a TT-tensor, say the TT-part of $\widehat{K}$, W is an unknown vector field to be determined, and $\mathbb{L}_g$ is the conformal Killing operator relative to g defined by (5.12).

If the right hand side of (5.12) vanishes, the vector field W is called a conformal Killing vector. By definition, any tensor of the form $\mathbb{L}_g Y$ for some vector field Y has trace free. The procedure of solving (5.11) is to find W and TT-part of $\widehat{K}$. This so-called TT-part is not unique in general and we have many ways to extract such a piece of information from $\widehat{K}$.

We first deal with W. In accordance with (5.16), we first have

$$(\widehat{K}_{TT})^{ij} = \widehat{K}^{ij} - (\mathbb{L}_g W)^{ij}. \tag{5.17}$$

The choice of the conformal Killing operator and the fact that $\widehat{K}$ is tracefree make the right hand side of (5.17) trace free. Besides, the transversality requirement $\nabla_i(\widehat{K}_{TT})^{ij} = 0$ and (5.11) lead to covariant equations for W given by

$$\nabla_i(\mathbb{L}_g W)^{ij} = \frac{n-1}{n} u^{\frac{2n}{n-2}} g^{ij} \partial_i \tau - \pi g^{ij} \partial_i \psi.$$

Using the convention $\Delta_{g,\mathbb{L}} = \mathrm{div}_g \circ \mathbb{L}_g$, we can rewrite the equation for W in the vector form as the following

$$\Delta_{g,\mathbb{L}} W = \frac{n-1}{n} u^{\frac{2n}{n-2}} \nabla_g \tau - \pi \nabla_g \psi. \qquad (5.18)$$

It is well-known that the operator $\Delta_{g,\mathbb{L}}$ which is similar to the vector Laplacian is a second order, self-adjoint, linear, elliptic operator whose kernel consists of the space of conformal Killing vector fields, see [CBru09, Appendix II]. Thus under some mild conditions, we can solve (5.18) for W up to conformal Killing vector fields. Notice that, any conformal Killing vector field does not constitute any extra information to $\widehat{K}_{TT}$ in (5.17).

We now consider the TT-tensor $\widehat{K}_{TT}$. The search of such a tensor is somewhat freely. Its procedure can be formulated as follows. We start with a freely chosen traceless 2-tensor Z, then we solve for Y from the following equation

$$\Delta_{g,\mathbb{L}} Y = -\mathrm{div}_g Z. \qquad (5.19)$$

The existence of some Y from (5.19) comes from Proposition 5.3 and the fact that $\mathrm{div}_g Z$ is orthogonal to the space of conformal Killing vector fields whose proof is just a simple application of integration by parts as follows

$$\int_M (\nabla_i Z^{ij}) H_j \, d\mathrm{vol}_g = -\int_M Z^{ij} (\mathbb{L}_g H)_{ij} \, d\mathrm{vol}_g = 0$$

where H is a conformal Killing vector field. Since

$$\nabla_j(\mathbb{L}_g Y + Z)^{ij} = (\Delta_{g,\mathbb{L}} Y)^i + \nabla_j Z^{ij} = 0$$

we know that the traceless tensor $\mathbb{L}_g Y + Z$ is also transverse, thus, a TT-tensor. Let us denote

$$\sigma = \mathbb{L}_g Y + Z. \qquad (5.20)$$

In conclusion, we first begin with a freely chosen Z and solve (5.19) for Y. We then solve (5.18) to find W. Finally, we find $\widehat{K} = \sigma + \mathbb{L}_g W$ by means of the decomposition (5.16). Such a $\widehat{K}$ will satisfy (5.11).

5.2.6. *Summary*

In conclusion, our transformed equations are the following:

$$-\frac{4(n-1)}{n-2}\Delta_g u + \left(\text{Scal}_g - |\nabla\psi|_g^2\right)u$$
$$= -\left(\frac{n-1}{n}\tau^2 - 2U(\psi)\right)u^{\frac{n+2}{n-2}} + (|\sigma + \mathbb{L}_g W|_g^2 + \pi^2)u^{-\frac{3n-2}{n-2}} \tag{5.21}$$

and

$$\Delta_{g,\mathbb{L}} W = \frac{n-1}{n}u^{\frac{2n}{n-2}}\nabla_g\tau - \pi\nabla_g\psi \tag{5.22}$$

where $\tau = \text{tr}_{\widetilde{g}} \widetilde{K}$. Clearly, the system (5.21)-(5.22) is coupling in the sense that we cannot solve each equation separately, not even in the vacuum case meaning that there is no π, ψ, and U. One way to decouple the system is to assume that the mean curvature τ is constant everywhere in M (hence we call CMC case for short).

5.3. *Transformed PDEs in the case $n = 2$*

In the previous section, we have shown how the conformal method may be used to derive the constraint equations with scalar fields when the dimension $n \geqslant 3$. In this section, we continue to use the conformal method to construct the constraint equations with scalar fields in the dimension $n = 2$. To the best of our knowledge, the first paper dealing with the Einstein equations in the two-dimensional cases is [Mon86]. Subsequently, some generalization of the corresponding situation for the Einstein–Maxwell and Einstein–Maxwell–Higgs equations were also obtained by the author in [Mon90]. Again, no result is known for the case of scalar fields.

5.3.1. *The momentum constraint equation*

In order to fix notations, we keep using $\widetilde{g} = e^{2\varphi}g$. We first prove the following simple result which was motivated by (5.5).

Lemma 5.4: *On a 2-dimensional manifold, if $\widetilde{g} = e^{2\varphi}g$, the covariant derivatives in $\widetilde{g}$ and g being respectively denoted $\widetilde{\nabla}$ and ∇, the divergences in the metrics $\widetilde{g}$ and g of an arbitrary contravariant 2-tensor P^{ij} verify the following identity*

$$\text{div}_{\widetilde{g}}P^{ij} = e^{-4\varphi}\text{div}_g(e^{4\varphi}P^{ij}) - g^{ij}\partial_i\varphi\,\text{tr}_g P. \tag{5.23}$$

Proof: The proof is almost similar to the proof of Lemma 5.2. Indeed, if we denote the tensor P by $P = P^{ij}\partial_i \otimes \partial_j$ then we first recall from (5.6) the following rule

$$\widetilde{\nabla}_i P^{ij} = \partial_i P^{ij} + P^{lj}\widetilde{\Gamma}^i_{li} + P^{il}\widetilde{\Gamma}^j_{li}.$$

Notice that under the conformal change $\widetilde{g} = e^{2\varphi}g$, the Christoffel symbols computed with respect to $\widetilde{g}$ and g now verify the following identity

$$\widetilde{\Gamma}^k_{ij} = \Gamma^k_{ij} + (\delta^k_i \partial_j\varphi + \delta^k_j \partial_i\varphi - g_{ij}g^{kl}\partial_l\varphi).$$

Therefore, thanks to $n = 2$

$$\begin{aligned}
\widetilde{\nabla}_i P^{ij} &= \partial_i P^{ij} + P^{lj}\Gamma^i_{li} + P^{il}\Gamma^j_{li} \\
&\quad + P^{lj}(\delta^i_l \partial_i\varphi + \delta^i_i \partial_l\varphi - g_{li}g^{im}\partial_m\varphi) \\
&\quad + P^{il}(\delta^j_l \partial_i\varphi + \delta^j_i \partial_l\varphi - g_{li}g^{jm}\partial_m\varphi) \\
&= \nabla_i P^{ij} + (P^{ij}\partial_i\varphi + 2P^{lj}\partial_l\varphi - P^{mj}\partial_m\varphi) \\
&\quad + (P^{ij}\partial_i\varphi + P^{jl}\partial_l\varphi - P^{il}g_{li}g^{jm}\partial_m\varphi) \\
&= \nabla_i P^{ij} + 4P^{ij}\partial_i\varphi - g^{jm}\partial_m\varphi\,\mathrm{tr}_g P \\
&= e^{-4\varphi}\nabla_i(e^{4\varphi}P^{ij}) - g^{jm}\partial_m\varphi\,\mathrm{tr}_g P.
\end{aligned}$$

The proof immediately follows.   □

Let us consider the momentum constraint. Using Lemma 5.23, for a 2-tensor $\widehat{K}$, we decompose it as follows

$$\widetilde{K}^{ij} = e^{-4\varphi}\widehat{K}^{ij} + \frac{\tau}{2}\widetilde{g}^{ij}.$$

This decomposition obeys the same properties as $n \geqslant 3$, that is, $\widehat{K}$ is symmetric and traceless. Therefore, by (5.23), one has the following

$$\begin{aligned}
\widetilde{J}^j + \widetilde{g}^{ij}\partial_i\tau &= \widetilde{\nabla}_i \widetilde{K}^{ij} \\
&= e^{-4\varphi}\nabla_i(e^{4\varphi}\widetilde{K}^{ij}) - g^{ij}\partial_i\varphi\,\mathrm{tr}_g \widetilde{K} \\
&= e^{-4\varphi}\nabla_i\left(\widehat{K}^{ij} + \frac{1}{2}e^{4\varphi}\widetilde{g}^{ij}\tau\right) - g^{ij}\partial_i\varphi\,\mathrm{tr}_g \widetilde{K} \\
&= e^{-4\varphi}\nabla_i\widehat{K}^{ij} + \frac{1}{2}e^{-4\varphi}\nabla_i\left(e^{2\varphi}g^{ij}\tau\right) - (e^{-2\varphi}g^{ij})\partial_i\varphi(\underbrace{e^{2\varphi}\,\mathrm{tr}_g \widetilde{K}}_{\tau}) \\
&= e^{-4\varphi}\nabla_i\widehat{K}^{ij} + \frac{1}{2}\underbrace{e^{-4\varphi}g^{ij}}_{e^{-2\varphi}\widetilde{g}^{ij}}\nabla_i(e^{2\varphi}\tau) - \widetilde{g}^{ij}\partial_i\varphi\tau \\
&= e^{-4\varphi}\nabla_i\widehat{K}^{ij} + \frac{1}{2}e^{-2\varphi}\widetilde{g}^{ij}\nabla_i\tau.
\end{aligned}$$

In other words,

$$\nabla_i \widehat{K}^{ij} = e^{4\varphi} \widetilde{J}^j + \frac{1}{2} e^{4\varphi} \widetilde{g}^{ij} \partial_i \tau.$$

Thus, the momentum constraint equation that we need is

$$\mathrm{div}_g \, \widehat{K} = \frac{1}{2} e^{2\varphi} \nabla_g \tau + e^{4\varphi} \widetilde{J}. \tag{5.24}$$

5.3.2. *The Hamiltonian constraint equation*

Let us now consider the Hamiltonian constraint. One first obtains by (5.3)

$$\mathrm{Scal}_{\widetilde{g}} = e^{-2\varphi} (\mathrm{Scal}_g - 2\Delta_g \varphi).$$

Consequently, the Hamiltonian constraint temporarily reads as the following

$$-2\Delta_g \varphi + \mathrm{Scal}_g - (|\widetilde{K}|^2_{\widetilde{g}} - \tau^2 + 2\overline{\rho}) e^{2\varphi} = 0.$$

Obviously,

$$|\widetilde{K}|^2_{\widetilde{g}} = \widetilde{g}_{ih} \widetilde{g}_{jk} \widetilde{K}^{ij} \widetilde{K}^{hk} = e^{-4\varphi} |\widehat{K}|^2_g + \frac{1}{2} \tau^2$$

which implies that

$$-2\Delta_g \varphi + \mathrm{Scal}_g - \left(e^{-4\varphi} |\widehat{K}|^2_g - \frac{1}{2} \tau^2 + 2\overline{\rho} \right) e^{2\varphi} = 0. \tag{5.25}$$

5.3.3. *Scaling of the scalar fields*

Using (5.9) one easily gets $N = e^{2\varphi} \widehat{N}$ which immediately implies

$$\widetilde{\pi} = N^{-1} \widetilde{\partial_0 \psi} = e^{-2\varphi} \pi.$$

Therefore, we can decompose the energy density on M as follows

$$\widetilde{\rho} = \frac{1}{2} (e^{-4\varphi} |\pi|^2 + e^{-2\varphi} |\nabla \psi|^2_{\widetilde{g}}) + U(\psi).$$

Now for the momentum density, one can see that

$$\widetilde{J}^i = -N^{-1} \widetilde{g}^{ij} \partial_j \psi \widetilde{\partial_0 \psi} = -e^{-4\varphi} g^{ij} \partial_j \psi \pi.$$

That is

$$\widetilde{J} = -e^{-4\varphi} \pi \nabla_g \psi.$$

Using the formula for $\widetilde{J}$, we can rewrite (5.24) as

$$\mathrm{div}_g \, \widehat{K} = \frac{1}{2} e^{2\varphi} \nabla_g \tau - \pi \nabla_g \psi. \tag{5.26}$$

We now regroup (5.25) as

$$-2\Delta_g\varphi + (\mathrm{Scal}_g - |\nabla\psi|_g^2) = -\left(\frac{1}{2}\tau^2 - 2U(\psi)\right)e^{2\varphi} + (|\pi|^2 + |\widehat{K}|_g^2)e^{-2\varphi}.$$
(5.27)

At this stage, the constraint equations in two dimensions become a system of partial differential equations (5.27)–(5.26). Notice that the coefficient of $e^{-2\varphi}$ is always non-negative in this setting. In the vacuum case, that is $U(\widetilde\psi) \equiv 0$, the coefficient of $e^{2\varphi}$ is always negative and one can use the method of sub- and super-solutions to prove the existence result, see [Mon86] for details.

6. Solving the Transformed PDEs in the Case $n \geqslant 3$: Mathematical Settings

Let us recall some information that we have already seen from previous sections. By using the conformal method we are able to transform the Einstein-scalar field constraint equations (4.1) and (4.2) into a couple of two equations given in (5.21)-(5.22). By denoting

$$\mathcal{R}_{g,\psi} = \frac{n-2}{4(n-1)}(\mathrm{Scal}_g - |\nabla\psi|_g^2),$$

$$\mathcal{B}_{\tau,\psi} = \frac{n-2}{4(n-1)}\left(\frac{n-1}{n}\tau^2 - 2U(\psi)\right),$$
(6.1)

$$\mathcal{A}_{g,W,\pi} = \frac{n-2}{4(n-1)}(|\sigma + \mathbb{L}_g W|_g^2 + \pi^2),$$

one can see that (5.21) simply becomes

$$-\Delta_g u + \mathcal{R}_{g,\psi} u = -\mathcal{B}_{\tau,\psi} u^{\frac{n+2}{n-2}} + \mathcal{A}_{g,W,\pi} u^{-\frac{3n-2}{n-2}}.$$
(6.2)

In order to make the study more general, we can consider a more general form of (6.2) which can be written as the following

$$-\Delta_g u + hu = fu^{\frac{n+2}{n-2}} + au^{-\frac{3n-2}{n-2}},$$
(6.3)

where h, f, and a are smooth functions. In this note, we also call (6.3) the Einstein-scalar field Lichnerowicz equation.

The main objective of the study is to determine which choices of the conformal data $(g, \sigma, \tau, \psi, \pi)$ permit one to solve Eqs. (5.21)-(5.22) for the determined data (u, W) and which do not. Clearly, Eq. (5.21) is a semilinear elliptic equation, called the Einstein-scalar field Lichnerowicz equation, for u provided σ and W are already known. However, be careful because Eq. (5.22) is a vector equation.

6.1. *Preliminaries*

6.1.1. *Sobolev spaces and related inequalities*

Given a smooth compact Riemannian manifold (M, g) of dimension n, one easily defines the Sobolev spaces $H_p^k(M)$ for any positive integers k and p. To be precise, we define $H_p^k(M)$ as the completion of $C^\infty(M)$ with respect to the following norm

$$\|u\|_{H_p^k} = \sum_{j=1}^{k} \|\nabla^j u\|_{L^p}.$$

When $p = 2$, we simply write $H_2^k(M)$ as $H^k(M)$.

By $\mathcal{K}_1$ and $\mathcal{A}_1$ we mean the positive constants such that the Sobolev inequality holds, that is, for all $u \in H^1(M)$,

$$\mathcal{K}_1 \int_M |\nabla u|^2 \, dvol_g + \mathcal{A}_1 \int_M u^2 \, dvol_g \geq \left(\int_M |u|^{\frac{2n}{n-2}} \, dvol_g \right)^{\frac{n-2}{n}}. \qquad (6.4)$$

We notice that those constants $\mathcal{K}_1$ and $\mathcal{A}_1$ are independent of u.

As one may observe from (6.3) that the operator appearing in the left hand side of (6.3), that is $-\Delta + h$, admits some interesting features when we impose some conditions on the potential h. A typical assumption that people usually make is to assume that $-\Delta + h$ is coercive. Roughly speaking, this is equivalent to saying that

$$\inf_{u \in H^1(M)} \frac{\int_M |\nabla u|^2 \, dvol_g + \int_M hu^2 \, dvol_g}{\int_M u^2 \, dvol_g} > 0.$$

In particular, one may see that

$$\|u\|_{H_h^1} = \left(\int_M |\nabla u|^2 \, dvol_g + \int_M hu^2 \, dvol_g \right)^{\frac{1}{2}}$$

is an equivalent norm in $H^1(M)$. It is standard to check that if $h > 0$ everywhere then $-\Delta + h$ is coercive.

Another useful inequality appearing in this setting is the following. For all $u \in H^1(M)$, there holds

$$\int_M |\nabla u|^2 \, dvol_g + \int_M hu^2 \, dvol_g \geq \mathcal{S}_h \left(\int_M |u|^{\frac{2n}{n-2}} \, dvol_g \right)^{\frac{n-2}{n}}, \qquad (6.5)$$

where the constant $\mathcal{S}_h$ is called the Sobolev constant and is independent of u.

6.1.2. *The Yamabe-scalar field conformal invariant*

Let us assume that (M, g) is a compact Riemannian manifold without boundary. We recall from the study of the Yamabe problem [Aub98, Tru68, Yam60] on (M, g) the conformal Laplacian operator $\mathcal{L}_g$ acting on a smooth function u is defined by

$$\mathcal{L}_g u = \Delta_g u - \frac{n-2}{4(n-1)} \operatorname{Scal}_g u. \tag{6.6}$$

Operator $\mathcal{L}_g$ has the conformal covariance property

$$\mathcal{L}_{\widetilde{g}} u = \theta^{-\frac{n+2}{n-2}} \mathcal{L}_g(\theta u) \tag{6.7}$$

for any $\widetilde{g} = \theta^{\frac{4}{n-2}} g$ for some $\theta > 0$ being a smooth function. Inspired by (6.6), the authors in [CBIP07] introduced the so-called conformal scalar field Laplacian operator $\mathcal{L}_{g,\psi}$ given by

$$\mathcal{L}_{g,\psi} u = \Delta_g u - \frac{n-2}{4(n-1)}(\operatorname{Scal}_g - |\nabla \psi|_g^2) u. \tag{6.8}$$

It follows from (6.6) and (6.8) that

$$\mathcal{L}_{g,\psi} u = \mathcal{L}_g u + \frac{n-2}{4(n-1)} |\nabla \psi|_g^2 u.$$

We wish that $\mathcal{L}_{g,\psi}$ also has the conformal covariance property. For this reason, we first have

$$|\nabla \psi|_{\widetilde{g}}^2 = \widetilde{g}^{ij} \partial_i \psi \partial_j \psi = \theta^{-\frac{4}{n-2}} g^{ij} \partial_i \psi \partial_j \psi = \theta^{-\frac{4}{n-2}} |\nabla \psi|_g^2.$$

This and (6.7) immediately give

$$\mathcal{L}_{\widetilde{g},\psi} u = \mathcal{L}_{\widetilde{g}} u + \frac{n-2}{4(n-1)} |\nabla \psi|_{\widetilde{g}}^2 u$$

$$= \theta^{-\frac{n+2}{n-2}} \mathcal{L}_g(\theta u) + \frac{n-2}{4(n-1)} \theta^{-\frac{4}{n-2}} |\nabla \psi|_g^2 u$$

$$= \theta^{-\frac{n+2}{n-2}} \mathcal{L}_g(\theta u) + \frac{n-2}{4(n-1)} \theta^{-\frac{n+2}{n-2}} |\nabla \psi|_g^2 (\theta u).$$

In other words, operator $L_{g,\psi}$ also verifies the same conformal covariance property (6.7) in the sense that

$$\mathcal{L}_{\widetilde{g},\psi} u = \theta^{-\frac{n+2}{n-2}} \mathcal{L}_{g,\psi}(\theta u). \tag{6.9}$$

We now define the so-called conformal-scalar field Dirichlet energy of u by

$$\mathcal{E}_{g,\psi}(u) = -\frac{4(n-1)}{n-2} \int_M u \mathcal{L}_{g,\psi} u \, d\mathrm{vol}_g$$

$$= -\frac{4(n-1)}{n-2} \int_M \left(|\nabla \psi|_g^2 + \frac{n-2}{4(n-1)} (\operatorname{Scal}_g - |\nabla \psi|_g^2) u^2 \right) d\mathrm{vol}_g$$

and the conformal-scalar field Sobolev quotient by

$$Q_{g,\psi}(u) = \frac{\mathscr{E}_{g,\psi}(u)}{\|u\|^2_{L^{\frac{2n}{n-2}}}}.$$

Using (6.9) one has

$$Q_{\widetilde{g},\psi}(u) = Q_{g,\psi}(\theta u) \tag{6.10}$$

where $\widetilde{g} = \theta^{\frac{4}{n-2}} g$. We denote by $[g]$ the conformal class of the metric g given by

$$[g] = \{\widetilde{g} = \theta^{\frac{4}{n-2}} g, \theta \in C^\infty(M), \theta > 0\}.$$

Then we define the so-called Yamabe-scalar field conformal invariant by

$$\mathcal{Y}_\psi([g]) = \inf_{u \in H^1(M)} Q_{g,\psi}(u). \tag{6.11}$$

By (6.10), it is obvious that $\mathcal{Y}_\psi([g])$ is independent of the choice of background metric g in the conformal class used to define it, and is therefore an invariant of the conformal class.

We observer from the Hölder inequality and the compactness of M that

$$\left| \int_M \left((\mathrm{Scal}_g - |\nabla \psi|^2_g) u^2 \right) \, d\mathrm{vol}_g \right| \leqslant C \|u\|^2_{L^{\frac{2n}{n-2}}}$$

for some positive constant C independent of u. Thus, $\mathcal{Y}_\psi([g])$ is finite. Consequently, by using the sign of $\mathcal{Y}_\psi([g])$, we may partition the set of pairs $([g], \psi)$ into three classes which we label $\mathcal{Y}^-$, $\mathcal{Y}^0$, $\mathcal{Y}^+$, and refer to as the negative, zero, and positive Yamabe-scalar field conformal invariants on M.

The following important result was proved in [CBIP07].

Proposition 6.1: *The following conditions are equivalent*

(i) $\mathcal{Y}_\psi([g]) > 0$ *(respectively $= 0$, < 0);*
(ii) *There exists a metric $\widetilde{g} \in [g]$ which satisfies*

$$\mathrm{Scal}_{\widetilde{g}} - |\nabla \psi|^2_{\widetilde{g}} > 0$$

everywhere on M (respectively $= 0$, < 0);
(iii) *For any metric $\widetilde{g} \in [g]$, the first eigenvalue λ_1 of the self-adjoint elliptic operator $-\mathcal{L}_{\widetilde{g},\psi}$ is positive (respectively zero, negative).*

For any constant $c > 0$ and any metric $\widetilde{g} \in [g]$, let us consider the following metric $\widehat{g} = c^{\frac{4}{n-2}}\widetilde{g}$. In terms of the metric g, one may write $\widehat{g} = (cu)^{\frac{4}{n-2}}g$ provided $\widetilde{g} = u^{\frac{4}{n-2}}g$. Then a direct computation shows that

$$\mathrm{Scal}_{\widehat{g}} - |\nabla\psi|_{\widehat{g}}^2 = \mathrm{Scal}_{\widehat{g}} - c^{-\frac{4}{n-2}}|\nabla\psi|_{\widetilde{g}}^2$$

$$= c^{-\frac{4}{n-2}}\mathrm{Scal}_{\widetilde{g}} - \frac{4(n-1)}{n-2}c^{-\frac{n+2}{n-2}}\Delta_{\widetilde{g}}(c) - c^{-\frac{4}{n-2}}|\nabla\psi|_{\widetilde{g}}^2$$

$$= c^{-\frac{4}{n-2}}\left(\mathrm{Scal}_{\widetilde{g}} - |\nabla\psi|_{\widetilde{g}}^2\right).$$

Therefore, we can extend the preceding proposition as follows.

Proposition 6.2: *The following conditions are equivalent:*

(i) $\mathcal{Y}_\psi([g]) > 0$ *(respectively $= 0$, < 0);*
(ii') *There exists a metric $\widetilde{g} \in [g]$ which satisfies*

$$\mathrm{Scal}_{c\widetilde{g}} - |\nabla\psi|_{c\widetilde{g}}^2 > 0$$

everywhere on M (respectively $= 0$, < 0) and for any constant $c > 0$;
(iii) *For any metric $\widetilde{g} \in [g]$, the first eigenvalue λ_1 of the self-adjoint elliptic operator $-\mathcal{L}_{\widetilde{g},\psi}$ is positive (respectively zero, negative).*

The advantage of Proposition 6.2 is that it allows us to assume that the manifold M has unit volume by choosing a suitable constant $c > 0$.

One of the most important property of Eq. (6.2) is that they are conformally covariant in the following sense, see [CBIP07].

Proposition 6.3: *Let $\mathcal{D} = (g, \sigma, \tau, \psi, \pi)$ be a conformal initial data set for the Einstein-scalar field constraint equations on M. If $\widetilde{g} = \theta^{\frac{4}{n-2}}g$ for a smooth positive function θ, then we define the corresponding conformally transformed initial data set by*

$$\widetilde{\mathcal{D}} = (\widetilde{g}, \widetilde{\sigma}, \widetilde{\tau}, \widetilde{\psi}, \widetilde{\pi}) = (\theta^{\frac{4}{n-2}}g, \theta^{-2}\sigma, \tau, \theta^{-\frac{2n}{n-2}}\psi, \pi).$$

Let W be the solution to the conformal form of the momentum constrain equation with respect to the conformal initial data set $\mathcal{D}$ (for which we assume that a solution exists), and let $\widetilde{W}$ be the solution of the momentum constrain equation with respect to the conformally transformed initial data set $\widetilde{\mathcal{D}}$ (which will exist if W does). Then u is a solution to Eq. (6.2) for the conformal data $\mathcal{D}$ with W if and only if $\theta^{-1}u$ is a solution to Eq. (6.2) for the transformed conformal data $\widetilde{\mathcal{D}}$ with $\widetilde{W}$.

Using Proposition 6.3 above, it turns out that the sign of $\mathcal{Y}_\psi([g])$ plays an important role in the study because we can first perform a conformal transformation on the conformal initial data from $(g, \sigma, \tau, \psi, \pi)$ to

$(\theta^{\frac{4}{n-2}}g, \theta^{-2}\sigma, \tau, \theta^{-\frac{2n}{n-2}}\psi, \pi)$ in such a way that $\mathcal{R}_{\tilde{g},\tilde{\psi}}$ has a fixed sign by means of Proposition 6.1. Therefore, it suffices to study the solvability of Eq. (6.2) for the transformed data $(\theta^{\frac{4}{n-2}}g, \theta^{-2}\sigma, \tau, \theta^{-\frac{2n}{n-2}}\psi, \pi)$ rather than to study to solvability of Eq. (6.2) for the original data $(g, \sigma, \tau, \psi, \pi)$.

6.2. *A classification of Choquet-Bruhat–Isenberg–Pollack*

While, as we have noted, the conformal method can be effectively applied for solving the constraint equations with scalar fields in most cases, it should be pointed out that there are several cases for which either partial result or no result was achieved, see [CBIP07] for details. In order to talk about the objectives of the current study, let us go back to [CBIP07]. A typical result contained in [CBIP07] is a fairly complete picture of sets of CMC conformal data which lead to solutions of the constraint equations with scalar fields and which do not. This picture can be summarized in the following two tables where

- 'Y' indicates that (6.2) can be solved for that class of conformal data;
- 'N' indicates that (6.2) has no positive solution;
- 'PR' indicates that we have partial results; and
- 'NR' indicates that for this class of initial data we have no results indicating existence or non-existence.

Table 1. Results for the case $\mathcal{A}_{\gamma,W,\pi} \equiv 0$.

$\mathcal{B}_{\tau,\psi}$	other	$\mathcal{B} < 0$	$\mathcal{B} \leqslant 0$	$\mathcal{B} \equiv 0$	$\mathcal{B} \geqslant 0$	$\mathcal{B} > 0$
$\mathcal{Y}_\psi < 0$	NR	N	N	N	PR	Y
$\mathcal{Y}_\psi = 0$	NR	N	N	Y	N	N
$\mathcal{Y}_\psi > 0$	PR	PR	PR	N	N	N

Table 2. Results for the case $\mathcal{A}_{\gamma,W,\pi} \not\equiv 0$.

$\mathcal{B}_{\tau,\psi}$	other	$\mathcal{B} < 0$	$\mathcal{B} \leqslant 0$	$\mathcal{B} \equiv 0$	$\mathcal{B} \geqslant 0$	$\mathcal{B} > 0$
$\mathcal{Y}_\psi < 0$	NR	N	N	N	PR	Y
$\mathcal{Y}_\psi = 0$	NR	N	N	N	Y	Y
$\mathcal{Y}_\psi > 0$	PR	PR	PR	Y	Y	Y

6.2.1. *Solving the momentum constraints*

As we have already known that the operator $\Delta_{g,\mathbb{L}}$ is a second-order, self-adjoint, linear, elliptic operator whose kernel consists of the space of conformal Killing vector fields, see [CBru09, Appendix II]. It follows from the Fredholm alternative that for a given set of functions (u, τ, ψ, π) we may solve the momentum constraint

$$\Delta_{g,\mathbb{L}} W = \frac{n-1}{n} u^{\frac{2n}{n-2}} \nabla_g \tau - \pi \nabla_g \psi$$

if either

- (M, g) admits no conformal Killing vector fields, and thus, W is unique;

or

- $\frac{n-1}{n} u^{\frac{2n}{n-2}} \nabla_g \tau - \pi \nabla_y \psi$ is orthogonal in the L^2 sense to the space of conformal Killing vector fields, see (5.15).

In the CMC case (hence $\frac{n-1}{n} u^{\frac{2n}{n-2}} \nabla_g \tau \equiv 0$) and (M, g) admits conformal Killing vector fields, it suffices to require that $\pi \nabla_g \psi$ is orthogonal to the space of conformal Killing vector fields as a consequence of (5.15). Notice that, under the CMC assumption, the momentum constraint equations consist of only scalar field ψ. Therefore, we can solely solve it to obtain W. Having the existence of W and the fact that our system of constraint equations are decoupled, we may solely solve the conformal form of the Hamiltonian constraint for u.

6.2.2. *Solving the Hamiltonian constraints*

Unlike the Einstein equations in the vacuum case where we know exactly which sets of CMC conformal data permit the corresponding Lichnerowicz equation to be solved and which do not by the seminal work by Isenberg [Ise95], the analysis for Eq. (6.2) is more complicated, primarily because there are more relevant possibilities for the signs of the coefficients in (5.21).

Let us recall from (6.2) that the corresponding Lichnerowicz equation is simply given by

$$-\Delta_g u + \mathcal{R}_{g,\psi} u = -\mathcal{B}_{\tau,\psi} u^{\frac{n+2}{n-2}} + \mathcal{A}_{g,W,\pi} u^{-\frac{3n-2}{n-2}}$$

where coefficients $\mathcal{R}_{g,\psi}$, $\mathcal{B}_{\tau,\psi}$, and $\mathcal{A}_{g,W,\pi}$ are given in (6.1). In [CBIP07], their classification only depends on the sign of $\mathcal{R}_{g,\psi}$ and $\mathcal{B}_{\tau,\psi}$ since $\mathcal{A}_{g,W,\pi} \geqslant 0$. As for $\mathcal{B}_{\tau,\psi}$, there are six different possibilities, namely, this coefficient can be strictly positive, greater than or equal to zero, identically zero, less than or equal to zero, strictly negative, or of changing sign. For $\mathcal{R}_{g,\psi}$, in view of Proposition 6.1, under a suitable conformal change, we can fix its sign, thus, there are three possibilities, namely, this could be negative, identically zero, or positive. These classification of sign combined with the two options $\mathcal{A}_{g,W,\pi} \equiv 0$ and $\mathcal{A}_{g,W,\pi} \not\equiv 0$ gives us a total of 36 classes of data, see Tables 1 and 2.

Based on this division, the authors in [CBIP07] proved for almost all cases, we do know which sets of data permit Eq. (6.2) to be solved and which do not. For a detailed statement of this result, we prefer to [CBIP07, Theorems 1 and 2].

6.3. The Lichnerowicz equations with $\mathcal{R}_{g,\psi}$ being constant

We now consider Eq. (6.3) as a slightly more general version of Eq. (6.2). By Proposition 6.1, we know that, after a suitable conformal transformation, the function h is a smooth function having a fixed sign on M. Again, by Proposition 6.1, the function h vanishes if we are in the null Yamabe-scalar field conformal invariant, that is equivalent to saying that h is constant (which is equal to zero) in this case.

In this subsection, we show that in fact if we are in the negative Yamabe-scalar field conformal invariant, namely $\mathcal{Y}_\psi([g]) < 0$, we still can perform another conformal transformation on the conformal data from $(g, \sigma, \tau, \psi, \pi)$ to $(u^{\frac{4}{n-2}} g, u^{-2}\sigma, \tau, u^{-\frac{2n}{n-2}}\psi, \pi)$ in such a way that h is a negative constant. Thanks to Proposition 6.3, we may freely assume that h is a negative constant on M.

Proposition 6.4: *There exists a smooth function $u > 0$ such that under the transformed data $(u^{\frac{4}{n-2}} g, u^{-2}\sigma, \tau, u^{-\frac{2n}{n-2}}\psi, \pi)$ obtained from data $(g, \sigma, \tau, \psi, \pi)$ through the conformal change $\tilde{g} = u^{\frac{4}{n-2}} g$, coefficient $\mathcal{R}_{\tilde{g},\tilde{\psi}}$ is a negative constant.*

Proof: First, in view of Proposition 6.1 we may assume that function $\mathcal{R}_{g,\psi} < 0$ in the original data, that is, $\mathrm{Scal}_g - |\nabla \psi|_g^2 < 0$. Notice that if

$\widetilde{g} = u^{\frac{4}{n-2}} g$, coefficient $\mathcal{R}_{\widetilde{g}, \widetilde{\psi}}$ verifies the following rule

$$\mathcal{R}_{\widetilde{g}, \widetilde{\psi}} = \frac{n-2}{4(n-1)} (\mathrm{Scal}_{\widetilde{g}} - u^{-\frac{4}{n-2}} |\nabla \psi|_g^2)$$

$$= \frac{n-2}{4(n-1)} \left(u^{-\frac{4}{n-2}} \mathrm{Scal}_g - \frac{4(n-1)}{n-2} u^{-\frac{n+2}{n-2}} \Delta_g u - u^{-\frac{4}{n-2}} |\nabla \psi|_g^2 \right)$$

$$= u^{-\frac{n+2}{n-2}} (-\Delta_g u + \mathcal{R}_{g, \psi} u),$$

$$(6.12)$$

which yields $-\Delta_g \varphi + \mathcal{R}_{g, \psi} \varphi = \mathcal{R}_{\widetilde{g}, \widetilde{\psi}} \varphi^{\frac{n+2}{n-2}}$. Therefore, in terms of our notation, it suffices to prove that the following equation with h and $\widetilde{h}$ being negative and $\widetilde{h}$ is constant

$$-\Delta_g u + hu = \widetilde{h} u^{\frac{n+2}{n-2}} \tag{6.13}$$

always admits a smooth positive solution u. We use the method of sub- and super-solutions to seal this issue.

Existence of a sub-solution. This is obvious since a sufficiently small, positive constant $\underline{u}$ will serve. To see this, one can choose any $\underline{u}$ satisfying the following $\underline{u}^{\frac{4}{n-2}} \leqslant (\sup_M h)/\widetilde{h}$. With this, we immediately have $-\Delta_g \underline{u} + h\underline{u} \leqslant \widetilde{h} \underline{u}^{\frac{n+2}{n-2}}$. By definition, $\underline{u}$ is a sub-solution of (6.13).

Existence of a super-solution. We also show that a sufficiently large positive constant $\overline{u}$ will serve. Indeed, similarly to the argument above, one can show that any positive constant $\overline{u}$ satisfying $\overline{u}^{\frac{4}{n-2}} \geqslant (\inf_M h)/\widetilde{h}$ is a super-solution to (6.13) in the sense that $-\Delta_g \overline{u} + h\overline{u} \geqslant \widetilde{h} \overline{u}^{\frac{n+2}{n-2}}$.

Finally, it is an easy task to select those $\underline{u}$ and $\overline{u}$ so that $\underline{u} < \overline{u}$. The method of sub- and super-solutions now guarantees that (6.13) admits a positive solution which turns out to be smooth by a simple regularity argument as in the study of the Yamabe problem. The proof is complete. $\square$

7. Solving the Transformed PDEs in the Closed Case when $n \geqslant 3$: The Case of Constant Mean Curvature

In this section, we are interested in the case when M is compact without boundary (i.e. closed). We also assume throughout the section that τ is constant everywhere in M; hence $\nabla_g \tau = 0$.

7.1. *The vacuum case*

As we are in the CMC case, the system (5.21)-(5.22) is decoupled; thus solving the constraint equations is reduced to solving the following equation

$$-\frac{4(n-1)}{n-2}\Delta_g u + \mathrm{Scal}_g\, u = -\frac{n-1}{n}\tau^2 u^{\frac{n+2}{n-2}} + |\sigma'|_g^2 u^{-\frac{3n-2}{n-2}}, \qquad (7.1)$$

where we simply write $\sigma' = \sigma + \mathbb{L}_g W$. In this context, Isenberg [Ise95] obtained a complete picture of the solvability of (7.1) as shown in Table 3 where:

- 'Y' indicates that (7.1) can be solved for that class of conformal data;
- 'N' indicates that (7.1) has no positive solution.

Table 3.   Results for the vacuum case in the case of constant mean curvature.

	$\sigma' \equiv 0, \tau = 0$	$\sigma' \equiv 0, \tau \neq 0$	$\sigma' \not\equiv 0, \tau = 0$	$\sigma' \not\equiv 0, \tau \neq 0$
$\mathcal{Y} < 0$	N	N	Y	Y
$\mathcal{Y} = 0$	Y	N	N	Y
$\mathcal{Y} > 0$	N	Y	N	Y

As emphasized in [BI04], the mean curvature of the data plays an important role in separating those sets of free data for which we know whenever a solution for Eq. (7.1) exists.

Let us now take a closer look into [Ise95]. In fact, the analysis in [Ise95] basically consists of the following three key steps:

- *The maximum principle*: Making use of the maximum principle is crucial in [Ise95] because this will help us to conclude whether the finding solution u is positive in M. In [Ise95, Appendix], Isenberg provided several versions of the maximum principle for the Laplacian on compact manifolds with or without boundary.
- *The Yamabe invariant $\mathcal{Y}$*: The use of the Yamabe invariant $\mathcal{Y}$ in [Ise95] is fundamental which helps to split the set of Riemannian metrics into different catalogue using the sign of its associated invariant. We note that the Yamabe invariant $\mathcal{Y}$ and the Yamabe-scalar field invariant $\mathcal{Y}_\psi$ introduced in (6.11) are different only by the presence of the field ψ; hence $\mathcal{Y}$ can be obtain from $\mathcal{Y}_\psi$ by getting rid of any ψ.
- *The method of sub- and super-solutions*: This is the main method used in [Ise95, Section 3]. The method is rather easy to implement which

roughly says that whenever we can find a pair of sub- and super-solutions $(\underline{u}, \overline{u})$ in the following sense

$$-\frac{4(n-1)}{n-2}\Delta_g \underline{u} + \mathrm{Scal}_g\, \underline{u} \leqslant -\frac{n-1}{n}\tau^2 \underline{u}^{\frac{n+2}{n-2}} + |\sigma'|_g^2 \underline{u}^{-\frac{3n-2}{n-2}}$$

and

$$-\frac{4(n-1)}{n-2}\Delta_g \overline{u} + \mathrm{Scal}_g\, \overline{u} \geqslant -\frac{n-1}{n}\tau^2 \overline{u}^{\frac{n+2}{n-2}} + |\sigma'|_g^2 \overline{u}^{-\frac{3n-2}{n-2}},$$

then we can conclude that (7.1) always admits at least one solution $u \in (\underline{u}, \overline{u})$ pointwise. Using this method, different sub- and super-solutions was found in [Ise95, Section 5], some are constant, some are the unique solution to some simple equations of the form $-\Delta u + u = c$.

It is worth noting that uniqueness property of solutions to (7.1) was also obtained in [Ise95, Section 6].

7.2. The case of scalar fields

In this subsection, instead of considering the vacuum case, we are interested in solving the constraint equations in the presence of real scalar fields. Since we still assume that τ is constant, solving these constraint equations is equivalent to solving Eq. (6.2) solely. However, in order to make our results available for a wider class of equations, we not only focus on Eq. (6.2) but also consider Eq. (6.3).

The following results corresponding to the three different sign of h basically come from our three joint papers with Xu in [NX12] for the negative h, in [NX15] for the vanishing h, and in [NX14] for the positive h. Our approach is variational and based on the so-called subcritical approach widely known when solving the Yamabe problem [AR73, Str08].

Recall that, generally, our Eq. (6.3) consists of the following three difficulties:

- a critical exponent,
- a sign-changing potential, and
- a negative exponent.

Therefore, from variational method point of view, we propose a unique method to tackle equations of the form (6.3) having all difficulties above at the same time.

7.2.1. *The case $h < 0$*

We are interested in the existence (if possible, the multiplicity and the uniqueness) of positive solutions to Eq. (6.2) in the negative Yamabe-scalar field conformal invariant, or more general, solutions to Eq. (6.3) in the case $h < 0$.

Notice that, as occurred in the well-studied in the prescribed scalar curvature problem, a simple use of integration by parts shows that the condition $\int_M f \, dvol_g < 0$ is also necessary in this case, see [NX12]. Perhaps, this condition reflects the condition $h < 0$ we are considering. However, there is another necessary condition in this case and this new necessary condition reflects the sign-changing property of f.

To state this new necessary condition, following [Rau95], we first define the following number

$$\lambda_f = \begin{cases} \displaystyle\inf_{u \in \mathscr{A}} \frac{\int_M |\nabla u|^2 \, dvol_g}{\int_M |u|^2 \, dvol_g}, & \text{if } \mathscr{A} \neq \emptyset, \\ +\infty, & \text{if } \mathscr{A} = \emptyset, \end{cases} \tag{7.2}$$

where

$$\mathscr{A} = \left\{ u \in H^1(M) : u \geqslant 0, \ u \not\equiv 0, \ \int_M |f^-| u \, dvol_g = 0 \right\} \tag{7.3}$$

and $f^+ = \max\{f, 0\}$ and $f^- = \min\{f, 0\}$. Clearly, functions in $\mathscr{A}$ are to be thought of as functions that vanish on the support of f^-. It is clear that $\lambda_f < +\infty$ if and only if the set $\{f \geqslant 0\}$ has positive measure, that is, either $\sup_M f > 0$ or $\sup_M f = 0$ and $\int_{\{f=0\}} 1 \, dvol_g > 0$. The second necessary condition that we would like to mention is $|h| < \lambda_f$, see [NX12]. As can be easily guessed, this condition shows us a strong connection between the sign of h and the size of f^+.

The basic content of our study consists of two main parts depending of the behavior of f. First, we consider the case when the function f takes both positive and negative values. Our main theorem for this part can be stated as follows.

Theorem 7.1: *Let (M, g) be a smooth compact Riemannian manifold without the boundary of dimension $n \geqslant 3$. Assume that f and $a \geqslant 0$ are smooth functions on M such that $\int_M f \, dvol_g < 0$, $\sup_M f > 0$, $\int_M a \, dvol_g > 0$, and $|h| < \lambda_f$ where λ_f is given in (7.2) below. Let us also suppose that the integral of a satisfies*

$$\int_M a \, dvol_g < \frac{1}{n-2} \left(\frac{n-1}{n-2}\right)^{n-1} \left(\frac{|h|}{\int_M |f^-| \, dvol_g}\right)^n \int_M |f^-| \, dvol_g \tag{7.4}$$

where f^- is the negative part of f. Then there exists a number $C > 0$ to be specified such that if

$$(\sup_M f) \left(\int_M |f^-| \, d\text{vol}_g \right)^{-1} < C, \qquad (7.5)$$

then Eq. (6.3) possesses at least two smooth positive solutions.

If we assume that f does not change sign in the sense that $f \leqslant 0$ in M, we obtain necessary and sufficient solvability conditions as pointed out in [CBIP07] in the case of (6.2). That is the content of the second part of our study for the case $h < 0$.

Theorem 7.2: *Let (M, g) be a smooth compact Riemannian manifold without boundary of dimension $n \geqslant 3$. Let $h < 0$ be a constant, f and a be smooth functions on M with $a \geqslant 0$ in M, $f \leqslant 0$ but not strictly negative. Then Eq. (6.3) possesses one positive solution if and only if $|h| < \lambda_f$.*

7.2.2. The case $h = 0$

Now we move to the next case, the case $h = 0$. We again continue to study some quantitative properties of positive, smooth solutions of Eq. (6.3) when $h = 0$.

In the previous subsection, we have already seen that, in the case $h < 0$, a suitable balance between coefficients h, f, a of (6.3) is enough to guarantee the existence of one positive smooth solution. In addition, it was found that under some further conditions we may or we may not have the uniqueness property of solutions of (6.3). This subsection is a continuation of the previous one where we consider the case when $h = 0$, that is, we are interested in the following simple partial differential equation

$$- \Delta_g u = f u^{\frac{n+2}{n-2}} + a u^{-\frac{3n-2}{n-2}}, \quad u > 0. \qquad (7.6)$$

The content of our study consists of three main parts. In the first part of the study, we mainly consider the case $\sup_M f > 0$. In this context, we was able to show that if $\sup_M f$ and $\int_M a \, d\text{vol}_g$ are small, then (7.6) possesses at least one smooth positive solutions. The first main theorem can be stated as follows.

Theorem 7.3: *Let (M, g) be a smooth compact Riemannian manifold without the boundary of dimension $n \geqslant 3$. Assume that f and $a \geqslant 0$ are smooth functions on M such that $\int_M a \, d\text{vol}_g > 0$, $\int_M f \, d\text{vol}_g < 0$, and $\sup_M f > 0$.*

Then there exist two positive numbers η_0 and λ depending only on the negative part f^- of f such that if

$$\sup_M f \leqslant \frac{\eta_0}{2} \int_M |f^-| \, d\mathrm{vol}_g \tag{7.7}$$

and

$$\int_M a \, d\mathrm{vol}_g < \frac{\lambda^n}{4(\eta_0)^{n-2}} \left(\frac{2n}{n-2}\right)^{n-2} \left(\int_M |f^-| \, d\mathrm{vol}_g\right)^{1-n} \tag{7.8}$$

hold, then Eq. (7.6) possesses at least one smooth positive solution.

By a simple comparison, one can easily see that except the multiplicity part, we was successful to carry the conclusion of Theorem 7.1 for the case $h < 0$ to the case $h = 0$. However, since Eq. (6.3) in the case $h = 0$ take a form simpler than that of the case $h < 0$, we might expect that the above two conditions (7.7) and (7.8) could be weakened.

Surprisingly, we was able to prove that the condition (7.7) can be relaxed. Unfortunately, for the price we pay, the estimate for $\int_M a \, d\mathrm{vol}_g$ needs to be replaced by another estimate for $\sup_M a$. This is the content of the second result.

Theorem 7.4: *Let (M, g) be a smooth compact Riemannian manifold without the boundary of dimension $n \geqslant 3$. Assume that f and $a \geqslant 0$ are smooth functions on M such that $\int_M a \, d\mathrm{vol}_g > 0$, $\int_M f \, d\mathrm{vol}_g < 0$, and $\sup_M f > 0$. Then if $\sup_M a$ is small, then Eq. (7.6) possesses at least one smooth positive solution.*

In the last part of the present study, we focus our attention to the case $\sup_M f \leqslant 0$. It should mention that in the statement of Theorem 7.3, $\sup_M f$ is nothing but $\sup_M f^+$ where f^+ is the positive part of f. Therefore, if we assume $f \leqslant 0$, we then see that the condition (7.7) is fulfilled for any small η_0. However, one can immediately observe that the right hand side of (7.8) goes to $+\infty$ as $\eta_0 \to 0$. This suggests that under the case $\sup_M f \leqslant 0$, there is no other condition for $\int_M a \, d\mathrm{vol}_g$ than $\int_M a \, d\mathrm{vol}_g > 0$. That is the content of our next result.

Theorem 7.5: *Let (M, g) be a smooth compact Riemannian manifold without boundary of dimension $n \geqslant 3$. Let f and a be smooth functions on M with $a \geqslant 0$ in M, $\int_M a \, d\mathrm{vol}_g > 0$, and $f \leqslant 0$. Then Eq. (7.6) always possesses one positive solution. In addition, this solution is unique.*

Concerning Theorems 7.2 and 7.5, it is worth noticing that these theorems generalize the same results obtained in [CBIP07] when our equation takes the form (6.2). Roughly speaking, it was proved in [CBIP07] by the method of sub- and super-solutions that (6.3) and (7.6) always possess one positive solution so long as the functions f and a take the form $f = -\mathcal{B}_{\tau,\psi}$ and $a = \mathcal{A}_{g,W,\pi}$ with $f \leqslant 0$ and $a \geqslant 0$. The main ingredient of the proof in [CBIP07] is the conformal invariant property of $\mathcal{B}_{\tau,\psi}$ and $\mathcal{A}_{g,W,\pi}$. Apparently, this property is no longer available in our general case.

7.2.3. The case $h > 0$

Finally, we consider the remaining case when $h > 0$. We continue our study of some quantitative properties of positive smooth solutions to Eq. (6.3) when $h > 0$.

As always, we assume hereafter that $f, h > 0$, and $a \geqslant 0$ are smooth functions on M with $\int_M a \, d\text{vol}_g > 0$. For the sake of simplicity, it is important to note that we can freely choose a background metric g such that manifold M has unit volume.

As far as we know, Eq. (6.3) with $h > 0$ was first considered in [HPP09] by using variational methods. In that elegant paper, Hebey–Pacard–Pollack proved, among other things, a fundamental existence result which roughly says that a suitable control of $\int_M a \, d\text{vol}_g$ from above is enough to guarantee the existence of one positive smooth solution. Their result basically makes use of the fact that the operator $-\Delta_g + h$ is coercive. Although the coerciveness property is slightly weaker than the condition $h > 0$, however as one can see from previous sections that this condition is enough to guarantee that $\|\cdot\|_{H^1_h}$ is an equivalent norm on $H^1(M)$. The advantage of this setting is that the first eigenvalue of the operator $-\Delta_g + h$ is strictly positive, and thus, various goods properties of the theory of weighted Sobolev spaces can be applied.

Using our notations, their result can be restated as follows: There exists a constant $C = C(n)$, $C > 0$ depending only on n, such that if

$$\|\varphi\|_{H^1_h}^{\frac{2n}{n-2}} \int_M \frac{a}{\varphi^{\frac{2n}{n-2}}} \, d\text{vol}_g \leqslant \frac{C}{(S_h \sup_M |f|)^{n-1}} \tag{7.9}$$

and

$$\int_M f \varphi^{\frac{2n}{n-2}} \, d\text{vol}_g > 0 \tag{7.10}$$

for some smooth function $\varphi > 0$ in M, then Eq. (6.3) possesses a smooth positive solution in the case $a > 0$[e].

As can be seen from (7.10), the condition $\sup_M f > 0$ is crucial. Therefore, it is not clear whether (6.3) possesses a smooth positive solution or not in the case $\sup_M f \leqslant 0$. Moreover, it is necessary to have $a > 0$ in M in order to get a positive lower bound for smooth solutions of (6.3). Besides, the condition (7.9) involves not only $\sup_M f^+$ but also $\inf_M f^-$. In other words, for given a, the negative part f^- of f cannot be too negative. This restriction basically reflects the fact that the energy functional has to verify the mountain pass geometry as their solution was found as a mountain pass point.

Our study for this case was also motivated by a recent paper [MW12]. In that paper, provided $\underline{u}$ is a positive smooth solution, Ma–Wei constructed a mountain pass solution of (6.3) of the form $\underline{u} + v$ for some smooth function $v > 0$. More precise, they proved that in the case $3 \leqslant n < 6$ and that the first eigenvalue of the following operator

$$-\Delta + h - \frac{n+2}{n-2} f \underline{u}^{\frac{4}{n-2}} + \frac{3n-2}{n-2} a \underline{u}^{\frac{4n-4}{n-2}} \tag{7.11}$$

is positive, (6.3) possesses a mountain pass, smooth, positive solution.

It is easy to see that the positivity of the first eigenvalue of the operator given in (7.11) immediately implies that the solution $\underline{u}$ is strictly stable. Therefore, it is natural to seek for positive smooth solutions of (6.3) as local minimizers.

Another reason that supports this approach is to look at the profile of the energy functional associated to (6.3). Due to the presence of the term $au^{-\frac{3n-2}{n-2}}$, the energy of u is very large when $\max_M u$ is small. Clearly, in the case $f \leqslant 0$, the energy of u is also large when $\max_M u$ is large. Consequently, a local minimizer of the energy functional should exist which could provide a possible solution. Similarly, if one assumes that $\sup_M f > 0$ and that the energy functional admits some mountain pass geometry, a local minimizer of the energy functional again exists.

While searching for positive smooth solutions of (6.3), we found that

---

[e]The upper bound for the integral of a appearing in (7.9) seems to be natural. This is because in the case $h > 0$, it was proved in [HPP09] that if $a \geqslant 0$, $f > 0$, and

$$\left(\frac{n^n}{(n-1)^{n-1}}\right)^{\frac{n+2}{4n}} \int_M a^{\frac{n+2}{4n}} f^{\frac{3n-2}{4n}} \, d\mathrm{vol}_g > \int_M h^{\frac{n+2}{4}} f^{\frac{2-n}{4}} \, d\mathrm{vol}_g,$$

then (6.3) does not admit any smooth positive solution.

the method used in the two cases $h \leqslant 0$ still works in this context. While the case $h \leqslant 0$ involves more conditions and our analysis of solvability of (6.3) strongly depends on the ratio between $\sup_M f$ and $\int_M |f^-| \, dvol_g$, the case $h > 0$ requires fewer conditions than the non-positive case. In fact, as we shall see later, in the case $\sup_M f > 0$, no condition for f is imposed and we are able to show that if $\int_M a \, dvol_g$ is small, then (6.3) possesses at least one smooth positive solutions since the condition for $\sup_M f$ can be absorbed to the condition for $\int_M a \, dvol_g$.

The first result can be stated as follows.

Theorem 7.6: *Let (M,g) be a smooth compact Riemannian manifold without the boundary of dimension $n \geqslant 3$. Assume that f, $h > 0$, and $a \geqslant 0$ are smooth functions on M such that $\int_M a \, dvol_g > 0$ and $\sup_M f > 0$. We assume further that there exists a constant*

$$\tau > \max \left\{ 1, \left(\frac{2}{S_h} \int_M h \, dvol_g \right)^{\frac{n}{n-2}} \right\}$$

such that

$$\int_M a \, dvol_g < \frac{(2n-1)^{n-1}}{2^{2n-1} n^n} \frac{S_h}{\tau} \left(\frac{S_h \tau}{\tau \sup_M f - \int_M f \, dvol_g} \right)^{n-1} \quad (7.12)$$

holds. Then (6.3) possesses at least one smooth positive solution.

Observe from (7.12) that τ plays no role but a scaling factor. Therefore, for given $\int_M a \, dvol_g$, we could select τ sufficiently large and $\sup_M f$ sufficiently small in such a way that (7.12) is fulfilled. This suggests that under the case when $\sup_M f$ is small, the condition for $\int_M a \, dvol_g$ appearing in (7.12) can be relaxed. We prove this affirmatively and that is the content of the following theorem.

Theorem 7.7: *Let (M,g) be a smooth compact Riemannian manifold without the boundary of dimension $n \geqslant 3$. Let f, h, and a be smooth functions on M with $h > 0$, $a \geqslant 0$ in M, $\int_M a \, dvol_g > 0$, and $\sup_M f > 0$. Then there exists a positive constant C to be specified later such that if $\sup_M f < C$, then Eq. (6.3) possesses one positive smooth solution.*

As always, we finally focus our attention to the case when $\sup_M f \leqslant 0$. In this context, we are able to get a complete characterization of the existence of solutions of (6.3) in the case when $f \leqslant 0$. Roughly speaking, it should mention that in the statement of Theorem 7.6, $\sup_M f$ is exactly $\sup_M f^+$ where f^+ is the positive part of f. Therefore, without any $\sup_M f$, one can immediately observe that the right hand side of (7.12) goes to $+\infty$ as $\tau \to +\infty$. This suggests that under the condition $\sup_M f \leqslant 0$, no condition is imposed.

Theorem 7.8: *Let (M,g) be a smooth compact Riemannian manifold without boundary of dimension $n \geqslant 3$. Let f, h, and a be smooth functions on M with $h > 0$, $a \geqslant 0$ in M, $\int_M a\, d\mathrm{vol}_g > 0$, and $f \leqslant 0$. Then Eq. (6.3) always possesses one and only one positive smooth solution.*

Concerning Theorem 7.8, it is worth noticing that it generalizes the same result obtained in [CBIP07] when our equation takes the form (6.2). Roughly speaking, it was proved in [CBIP07] by the method of sub- and super-solutions that (6.3) always possesses one positive solution so long as the functions f and a take the form $f = -\mathcal{B}_{\tau,\psi}$ and $a = \mathcal{A}_{g,W,\pi}$ with $f \leqslant 0$ and $a \geqslant 0$. The main ingredient of the proof in [CBIP07] is the conformal invariant property of $\mathcal{B}_{\tau,\psi}$ and $\mathcal{A}_{g,W,\pi}$. Apparently, this property is no longer available in our general case.

7.2.4. *Summary*

Finally, before closing the present subsection, we would like to mention the interaction between the coefficients of the Einstein-scalar field Lichnerowicz equations (6.3) for arbitrary sign of h.

Using our results for the negative case in [NX12], for the null case in [NX15], and for the positive case in [NX14], one can obtain the following table which shows us how the coefficients in (6.3) depend on each other in order to get the existence of solutions.

Table 4. Interaction between the coefficients of (6.3) for any h.

h	$-\lambda_f < h < 0$	$h = 0$	$h = 0$	$h > 0$	$h > 0$	$h \gg 0$
f	$\dfrac{\sup f^+}{C(f^-)}$ <	$\dfrac{\sup f^+}{C(f^-)}$ <			$\dfrac{\sup f^+}{C(f^-,a)}$ <	
a	$\int a < C(f^-)$	$\dfrac{\int a}{C(f^-)}$ <	$\dfrac{\sup a}{C(f)}$ <	$\int a < C(f)$		

The interpretation of Table 4 is as follows: The second column basically says that h cannot be too negative as it must satisfy $h > -\lambda_f$ for some positive constant λ_f given in (7.2). Under this condition, we guarantee an existence result for (6.3) provided $\sup_M f^+$ and $\int_M a\, d\mathrm{vol}_g$ are bounded in terms of f^-. This result still holds for the case $h = 0$; however, the boundedness of $\sup_M f^+$ can be relaxed if we replace $\int_M a \, dvol_g$ by $\sup_M a$ as shown in the third column.

The fourth column shows that in the case $h > 0$, the boundedness of $\sup_M a$ can be weakened by using $\int_M a \, dvol_g$. For the fifth column, it shows that no condition is required if $\sup_M f^+$ is small in terms of f^- and a.

In the last column, it shows that (6.3) is always solvable if $\inf_M h$ is sufficiently large, for example, if h satisfies

$$h \geqslant \sup_M f + \sup_M a$$

in M. This is because in this case the constant 1 is a super-solution for (6.3) and this is enough since a sub-solution for (6.3) which is less than 1 always exists.

Finally, it is worth mentioning that our approach is quite useful as it can be used to tackle other problems, for example, see [NZ15].

7.3. Strategy for finding solutions and ideas of proofs

As we have noted before, in order to study (6.3), we follow the subcritical approach, i.e. as a standard routine, in the first step to tackle (6.3), we look for positive smooth solutions of the following subcritical problem

$$- \Delta_g u + hu = f|u|^{q-2}u + \frac{au}{(u^2 + \varepsilon)^{\frac{q}{2}+1}}, \tag{7.13}$$

where $\varepsilon > 0$ is small and $q \in (2, \frac{2n}{n-2})$ is close to $\frac{2n}{n-2}$.

Since the subcritical equation (7.13) is variational, we study (7.13) by using the variational method [AR73, Str08]. Our main procedure is to show that solutions of (6.3) exist as first $\varepsilon \searrow 0$ and then $q \nearrow \frac{2n}{n-2}$ under various assumptions. The whole procedure can be summarized as the following diagram.

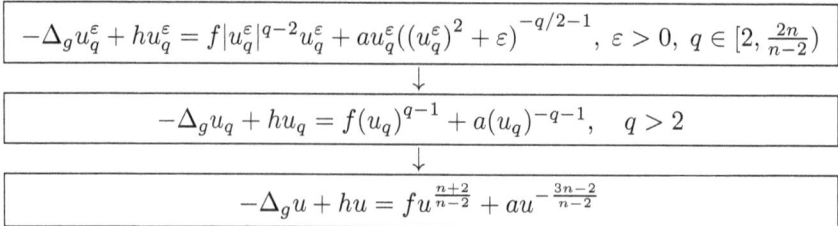

$$\boxed{-\Delta_g u_q^\varepsilon + h u_q^\varepsilon = f|u_q^\varepsilon|^{q-2} u_q^\varepsilon + a u_q^\varepsilon ((u_q^\varepsilon)^2 + \varepsilon)^{-q/2-1}, \quad \varepsilon > 0, \ q \in [2, \tfrac{2n}{n-2})}$$

$$\downarrow$$

$$\boxed{-\Delta_g u_q + h u_q = f(u_q)^{q-1} + a(u_q)^{-q-1}, \quad q > 2}$$

$$\downarrow$$

$$\boxed{-\Delta_g u + hu = f u^{\frac{n+2}{n-2}} + a u^{-\frac{3n-2}{n-2}}}$$

It is worth noticing that in [HPP09], the authors just considered (7.13) with q replaced by $\frac{2n}{n-2}$ and they found solutions as mountain pass points.

This main deference somehow reflects the fact that we need the compact embedding $H^1(M) \hookrightarrow L^q(M)$ while searching for minimum points.

In the following part of this subsection, we consider necessary conditions for f and h in the case $h \leqslant 0$.

7.3.1. *Necessary conditions for f and h in the case $h \leqslant 0$*

First, we derive a condition for $\int_M f \, dvol_g$ so that (7.13) admits positive smooth solutions. Our argument was motivated from a same result for the well-known prescribing scalar curvature problem. It is important to note that in the case $h > 0$, there is no such a condition on $\int_M f \, dvol_g$.

Proposition 7.9: *Assume that $h \leqslant 0$ is constant. Then then necessary condition for f so that Eq. (7.13) admits positive smooth solution is $\int_M f \, dvol_g < 0$. In particular, the necessary condition for (6.3) to have positive smooth solution is $\int_M f \, dvol_g < 0$.*

Proof: We assume that $u > 0$ is a smooth solution of (7.13). By multiplying both sides of (7.13) by u^{1-q}, one gets

$$(-\Delta_g u)u^{1-q} + hu^{3-q} = f + \frac{au^{2-q}}{(u^2 + \varepsilon)^{\frac{q}{2}+1}}.$$

Integrating over M and noticing that $h \leqslant 0$ give

$$\int_M (-\Delta_g u)u^{1-q} \, dvol_g > \int_M f \, dvol_g + \int_M \frac{au^{2-q}}{(u^2 + \varepsilon)^{\frac{q}{2}+1}} \, dvol_g.$$

By the divergence theorem, one obtains

$$\int_M (-\Delta_g u)u^{1-q} \, dvol_g = (1-q) \int_M u^{-q}|\nabla u|^2 \, dvol_g.$$

This and the fact that $q > 2$ deduce that

$$\int_M f \, dvol_g + \int_M \frac{au^{2-q}}{(u^2 + \varepsilon)^{\frac{q}{2}+1}} \, dvol_g < 0.$$

Obviously, there holds $\int_M f \, dvol_g < 0$ as claimed.   □

We are now study necessary conditions for h. More precisely, we show that the condition $|h| < \lambda_f$ is necessary if $\lambda_f < +\infty$ in order for (6.3) to have a positive smooth solution. Here we provide a different proof which is shorter than the proof in [Rau95, Section III.3]. Our argument depends on a Picone type identity for integrals [AP03] whose proof makes use of the density.

Lemma 7.10: *Assume that $v \in H^1(M)$ with $v \geqslant 0$ and $v \not\equiv 0$. Suppose that $u > 0$ is a smooth function. Then we have*

$$\int_M |\nabla v|^2 \, \mathrm{dvol}_g = -\int_M \frac{\Delta u}{u} v^2 \, \mathrm{dvol}_g + \int_M u^2 \left|\nabla \left(\frac{v}{u}\right)\right|^2 \mathrm{dvol}_g.$$

Proof: By density, there exist a family of regular functions $\{v_j\}_j$ such that

$$v_j \to v \text{ strongly in } H^1(M), \quad v_j \to v \text{ a.e. in } M, \quad v_j \in C^1(M).$$

The standard Picone identity tells us that

$$\int_M |\nabla v_j|^2 \, \mathrm{dvol}_g = -\int_M \frac{\Delta u}{u} v_j^2 \, \mathrm{dvol}_g + \int_M u^2 \left|\nabla \left(\frac{v_j}{u}\right)\right|^2 \mathrm{dvol}_g. \quad (7.14)$$

Since $u > 0$ is smooth and $v_j \to v$ strongly in $H^1(M)$ we immediately have

$$\int_M |\nabla v_j|^2 \, \mathrm{dvol}_g \to \int_M |\nabla v|^2 \, \mathrm{dvol}_g$$

and

$$\int_M \frac{\Delta u}{u} v_j^2 \, \mathrm{dvol}_g \to \int_M \frac{\Delta u}{u} v^2 \, \mathrm{dvol}_g$$

as $j \to +\infty$. Again using the smoothness of u, we can check that $v_j/u \to v/u$ strongly in $H^1(M)$. Therefore,

$$\int_M u^2 \left|\nabla \left(\frac{v_j}{u}\right)\right|^2 \mathrm{dvol}_g \to \int_M u^2 \left|\nabla \left(\frac{v}{u}\right)\right|^2 \mathrm{dvol}_g$$

as $j \to +\infty$. The proof now follows by taking the limit in (7.14) as $j \to +\infty$.

$\square$

We are now in a position to provide a different proof for necessary condition $|h| < \lambda_f$.

Proposition 7.11: *If Eq. (6.3) has a positive smooth solution, it is necessary to have $|h| < \lambda_f$.*

Proof: We only need to consider the case $\lambda_f < +\infty$ since otherwise it is trivial. We let $v \in \mathscr{A}$ arbitrary and assume that u is a positive smooth solution to (6.3). Using Lemma 7.10 and (6.3), we find that

$$\int_M |\nabla v|^2 \, dvol_g = |h| \int_M v^2 \, dvol_g + \int_M f u^{\frac{4}{n-2}} v^2 \, dvol_g$$
$$+ \int_M a v^2 u^{\frac{4-4n}{n-2}} \, dvol_g + \int_M u^2 \left| \nabla \left(\frac{v}{u} \right) \right|^2 \, dvol_g$$
$$\geqslant |h| \int_M v^2 \, dvol_g + \int_M u^2 \left| \nabla \left(\frac{v}{u} \right) \right|^2 \, dvol_g.$$

In other words, there holds

$$\frac{\int_M |\nabla v|^2 \, dvol_g}{\int_M v^2 \, dvol_g} \geqslant |h| + \frac{\int_M u^2 \left| \nabla \left(\frac{v}{u} \right) \right|^2 \, dvol_g}{\int_M v^2 \, dvol_g}. \tag{7.15}$$

In particular, $\lambda_f \geqslant |h| > 0$ by taking the infimum with respect to v. Notice that

$$\frac{\int_M u^2 \left| \nabla \left(\frac{v}{u} \right) \right|^2 \, dvol_g}{\int_M v^2 \, dvol_g} \geqslant \left(\frac{\inf_M u}{\sup_M u} \right)^2 \frac{\int_M \left| \nabla \left(\frac{v}{u} \right) \right|^2 \, dvol_g}{\int_M \left(\frac{v}{u} \right)^2 \, dvol_g} \geqslant \lambda_f \left(\frac{\inf_M u}{\sup_M u} \right)^2$$

since $v/u \in \mathscr{A}$. Having this, we can check from (7.15) that

$$\frac{\int_M |\nabla v|^2 \, dvol_g}{\int_M v^2 \, dvol_g} \geqslant |h| + \lambda_f \left(\frac{\inf_M u}{\sup_M u} \right)^2.$$

By taking the infimum with respect to v, we obtain

$$\lambda_f \geqslant |h| + \lambda_f \left(\frac{\inf_M u}{\sup_M u} \right)^2.$$

This and the fact that $\lambda > 0$ give us the desired result. □

From the above proof, one can observe that the function a plays no role but $a \geqslant 0$.

7.3.2. *Lower bounds for solutions of (7.13)*

We note that when sending $\varepsilon \searrow 0$ to get a solution u_q from a sequence of solutions $\{u_q^\varepsilon\}_\varepsilon$ to (7.13), one needs to take care the singularity by forcing u_q^ε to stay away from 0.

In the case $h < 0$, one can show that positive C^2 solutions to (7.13) is indeed bounded from below and away from zero.

Lemma 7.12: *Let u be a positive C^2 solution of (7.13) and assume that $q \in \left(\frac{2n-2}{n-2}, \frac{2n}{n-2} \right)$ is arbitrary. Then:*

- *If $h < 0$ is constant, then there holds*

$$\min_M u \geqslant \min\left\{\left(\frac{h}{\inf_M f}\right)^{\frac{n-2}{2}}, 1\right\} > 0 \qquad (7.16)$$

for any $\varepsilon > 0$.
- *If $h \equiv 0$, then there holds*

$$\min_M u \geqslant \frac{1}{2}\min\left\{\left(\frac{\inf_M a}{-\inf_M f}\right)^{\frac{n-2}{4n-4}}, 1\right\} \qquad (7.17)$$

for any

$$\varepsilon < \frac{1}{2}\min\left\{\left(\frac{\inf_M a}{-\inf_M f}\right)^{\frac{n-2}{2n-2}}, 1\right\}. \qquad (7.18)$$

- *If $h > 0$, then there holds*

$$\min_M u \geqslant \min\left\{\left(2^{-\frac{2n-2}{n-2}}\frac{\inf_M a}{\sup_M h + \sup_M |f|}\right)^{\frac{n-2}{4n-6}}, 1\right\} \qquad (7.19)$$

for any

$$\varepsilon < \min\left\{\left(2^{-\frac{2n-2}{n-2}}\frac{\inf_M a}{\sup_M h + \sup_M |f|}\right)^{\frac{2n-4}{4n-6}}, 1\right\}. \qquad (7.20)$$

Proof: Note that in the case $h > 0$, such a result was proved in [HPP09], we then omit the proof and focus only on the case $h \leqslant 0$. The only difference is that we calculate a precise lower bound for solutions of (7.13) as shown in (7.19), a detailed computation can be found in [Ngo12, Lemma 3.3].

In the case that $h < 0$ is constant, we first assume that u achieves its minimum value at x_0. For the sake of simplicity, we denote $u(x_0)$, $f(x_0)$, and $a(x_0)$ by u_0, f_0, and a_0 respectively. Notice that $u_0 > 0$ since u is a positive solution. We then have $\Delta_g u|_{x_0} \geqslant 0$; in particular,

$$hu_0 \geqslant f_0(u_0)^{q-1} + \frac{a_0 u_0}{((u_0)^2 + \varepsilon)^{\frac{q}{2}+1}} \geqslant f_0(u_0)^{q-1}.$$

Consequently, we get $f_0 < 0$ and thus $0 < \frac{h}{f_0} \leqslant (u_0)^{q-2}$ which immediately implies

$$\min_M u \geqslant \left(\frac{h}{\inf_M f}\right)^{\frac{1}{q-2}} \geqslant \min\left\{\left(\frac{h}{\inf_M f}\right)^{\frac{n-2}{2}}, 1\right\}$$

for any $q \in (\frac{2n-2}{n-2}, \frac{2n}{n-2})$ and any $\varepsilon > 0$. This proves the estimate (7.16) in our lemma.

Next we consider the case $h \geqslant 0$. Unlike the case $h < 0$ where we assume no condition on a, for the case $h \geqslant 0$, we do require $\inf_M a > 0$ since the lower bound for any positive C^2 solution u depends on $\inf_M a > 0$ as we can see from (7.17). Also, we require ε to be small as indicated in (7.18).

Now, let us assume that u achieves its minimum value at x_0. For simplicity, let us denote $u(x_0)$, $f(x_0)$, and $a(x_0)$ by u_0, f_0, and a_0 respectively. Notice that $u_0 > 0$ since u is a positive solution. We then have $\Delta_g u|_{x_0} \geqslant 0$; in particular,

$$f_0(u_0)^{q-1} + \frac{a_0 u_0}{((u_0)^2 + \varepsilon)^{\frac{q}{2}+1}} \leqslant 0. \tag{7.21}$$

Consequently, we get that $f_0 < 0$. Using (7.21) we can see that

$$a_0 \leqslant -f_0(u_0)^{q-2}((u_0)^2 + \varepsilon)^{\frac{q}{2}+1} \leqslant -f_0((u_0)^2 + \varepsilon)^q$$

which implies that

$$(u_0)^2 + \varepsilon \geqslant \left(\frac{a_0}{-f_0}\right)^{\frac{1}{q}} \geqslant \left(\frac{\inf_M a}{-\inf_M f}\right)^{\frac{1}{q}}.$$

Thus, one can conclude that u_0 satisfies (7.17) for any $q \in [\frac{2n-2}{n-2}, \frac{2n}{n-2})$ and any ε verifying the condition (7.18). The proof is complete. □

7.3.3. *Regularity for non-negative weak solutions of (7.13)*

This subsection is devoted to the regularity of weak solutions of (7.13). Despite the fact that h can be chosen as a constant in the non-positive Yamabe-scalar field conformal invariant, in this subsection, we allow h to be non-constant. As such, we assume that h, f and $a \geqslant 0$ are smooth functions and that the function h has a fixed sign on M.

Lemma 7.13: *Assume that $u \in H^1(M)$ is an almost everywhere non-negative weak solution of Eq. (7.13). We assume further that $\inf_M a > 0$ in the case when $h \geqslant 0$. Then*

(a) If $\varepsilon > 0$, then $u \in C^\infty(M)$. In particular, $u \geqslant 0$ in M.
(b) If $\varepsilon = 0$ and $u^{-1} \in L^p(M)$ for all $p \geqslant 1$, then $u \in C^\infty(M)$.

Proof: We first rewrite (7.13) as

$$-\Delta_g u + b(x)(1 + u) = 0$$

with

$$b(x) = \frac{u(x)}{1 + u(x)} \left(h(x) - \frac{a(x)}{(u(x)^2 + \varepsilon)^{\frac{q}{2}+1}} - f(x)|u(x)|^{q-2} \right).$$

By the Sobolev embedding, we know that $u \in L^q(M)$ for any $q \in (\frac{2n-2}{n-2}, \frac{2n}{n-2}]$. This and the conditions in both cases (a) and (b) imply

$$h(x) - \frac{a}{(u^2 + \varepsilon)^{\frac{q}{2}+1}} - f|u|^{q-2} \in L^{\frac{q}{q-2}}(M).$$

Notice that from $q \leqslant \frac{2n}{n-2}$ there holds $\frac{q}{q-2} \geqslant \frac{n}{2}$. We now use the Brezis–Kato estimate [Str08, Lemma B.3] to conclude that $u \in L^s(M)$ for any $s > 0$. Thus the Caldéron–Zygmund inequality implies that $u \in H^p(M)$ for any $p > 1$. The Sobolev embedding again implies that u is in $C^{0,\alpha}(M)$ for some $\alpha \in (0, 1)$. Thus, we know from the Schauder theory that $u \in C^{2,\alpha}(M)$ for some $\alpha \in (0, 1)$. In particular, u has a strictly positive lower bound by means of Lemma 7.12. Since u stays away from zero, we can iterate this process to conclude $u \in C^\infty(M)$. $\qquad\square$

7.3.4. Analysis of the energy functional

For each $q \in [\frac{2n-2}{n-2}, \frac{2n}{n-2})$ and $k > 0$, we introduce $\mathscr{B}_{k,q}$, a hyper-surface of $H^1(M)$, which is defined by

$$\mathscr{B}_{k,q} = \left\{ u \in H^1(M) : \|u\|_{L^q}^q = k \right\}. \tag{7.22}$$

Notice that for any $k > 0$, our set $\mathscr{B}_{k,q}$ is non-empty since it always contains $k^{\frac{1}{q}}$. Now we construct the energy functional associated to problem (7.13). For each $\varepsilon > 0$, we consider the functional $F_q^\varepsilon : H^1(M) \to \mathbb{R}$ defined by

$$F_q^\varepsilon(u) = \frac{1}{2} \int_M |\nabla u|^2 \, dvol_g + \frac{h}{2} \int_M u^2 \, dvol_g$$
$$- \frac{1}{q} \int_M f|u|^q \, dvol_g + \frac{1}{q} \int_M \frac{a}{(u^2 + \varepsilon)^{\frac{q}{2}}} \, dvol_g.$$

By a fairy standard argument, F_q^ε is continuously differentiable on $H^1(M)$, see Lemma 7.14 below, and thus weak solutions of (7.13) correspond to critical points of the functional F_q^ε. Now we set

$$\mu_{k,q}^\varepsilon = \inf_{u \in \mathscr{B}_{k,q}} F_q^\varepsilon(u).$$

By the Hölder inequality, it is not hard to see that $F_q^\varepsilon|_{\mathscr{B}_{k,q}}$ is bounded from below by $-k \sup_M f + \frac{h}{2} k^{\frac{2}{q}}$ and thus $\mu_{k,q}^\varepsilon > -\infty$ if k is finite. On the other hand, using the test function $u = k^{\frac{1}{q}}$, we get

$$\mu_{k,q}^\varepsilon \le \frac{h}{2} k^{\frac{2}{q}} - \frac{k}{q} \int_M f \, d\text{vol}_g + \frac{1}{q} \int_M \frac{a}{\left(k^{\frac{2}{q}} + \varepsilon\right)^2} \, d\text{vol}_g \qquad (7.23)$$

which tells us that $\mu_{k,q}^\varepsilon < +\infty$. Our aim was to find critical points of the functional F_q^ε. In order to support our aim, we prove the following result.

Lemma 7.14: *The first variation of F_q^ε at a point u in a direction v is given by*

$$\delta F_q^\varepsilon(u)(v) = \int_M \nabla u \cdot \nabla v \, d\text{vol}_g + h \int_M uv \, d\text{vol}_g$$
$$- \int_M f|u|^{q-2} uv \, d\text{vol}_g - \int_M \frac{au}{(u^2 + \varepsilon)^{\frac{q}{2}+1}} v \, d\text{vol}_g$$

where $\nabla u \cdot \nabla v$ stands for the pointwise scalar product of ∇u and ∇v with respect to the metric g.

Proof: The proof is simple, we include it here for completeness. In fact, for any smooth function v, there holds

$$\delta F_q^\varepsilon(u)(v) = \frac{d}{dt} F_q^\varepsilon(u + tv) \Big|_{t=0}$$
$$= \frac{d}{dt} \left[\frac{1}{2} \int_M |\nabla(u + tv)|^2 \, d\text{vol}_g + \frac{h}{2} \int_M (u + tv)^2 \, d\text{vol}_g \right] \Big|_{t=0}$$
$$+ \frac{d}{dt} \left[-\frac{1}{q} \int_M f|u + tv|^q \, d\text{vol}_g \right] \Big|_{t=0}$$
$$+ \frac{d}{dt} \left[\frac{1}{q} \int_M a((u + tv)^2 + \varepsilon)^{-\frac{q}{2}} \, d\text{vol}_g \right] \Big|_{t=0}$$
$$= \int_M (-\Delta_g u)v \, d\text{vol}_g + h \int_M uv \, d\text{vol}_g$$
$$- \int_M f|u|^{q-2} uv \, d\text{vol}_g - \int_M auv(u^2 + \varepsilon)^{-\frac{q}{2}-1} \, d\text{vol}_g,$$

which provides the desired result. $\square$

An interesting fact due to the subcritical approach is that $\mu_{k,q}^\varepsilon$ is achieved for each k, q, and ε fixed.

Lemma 7.15: *For each $k > 0$, $q \in (2, \frac{2n}{n-2})$, and $\varepsilon > 0$ fixed, the number $\mu_{k,q}^\varepsilon$ is achieved by some smooth positive function.*

The proof is standard and is based on the so-called direct methods in the calculus of variations, see [Ngo12, Lemma 4.2]. Another interesting feature of $\mu_{k,q}^{\varepsilon}$ is the continuity of $\mu_{k,q}^{\varepsilon}$ with respect to k.

Lemma 7.16: *For $\varepsilon > 0$ fixed, $\mu_{k,q}^{\varepsilon}$ is continuous with respect to k.*

The proof of Lemma 7.16 is also standard, see [Ngo12, Proposition 4.1]. The key idea is to make of the weakly lower semi-continuity of F_q^{ε} and the fact that $q < \frac{2n}{n-2}$.

The next part of our analysis for F_q^{ε} is to study the asymptotic behavior of $\mu_{k,q}^{\varepsilon}$ when k varies. Depending on the sign of $\sup_M f$, the behavior of $\mu_{k,q}^{\varepsilon}$ varies. In the dedicated case when $\sup_M f > 0$ and $\inf_M < 0$, i.e. when f changes sign, we can prove that when k is small, $\mu_{k,q}^{\varepsilon}$ goes to $+\infty$ and when k is big, $\mu_{k,q}^{\varepsilon}$ goes to $-\infty$. This can be plotted as in Fig. 4.

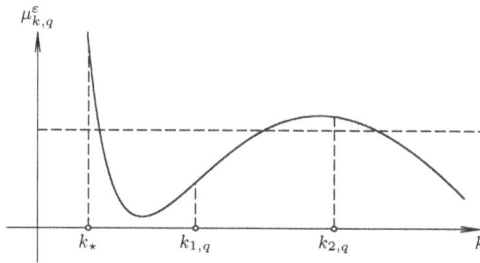

Fig. 4.   The asymptotic behavior of $\mu_{k,q}^{\varepsilon}$ when $\sup_M f > 0$.

This is the content of the following two lemmas.

Lemma 7.17: *There holds $\lim_{k \to 0+} \mu_{k,q}^{k^{\frac{2}{q}}} = +\infty$. In particular, there is some $k_\star$ sufficiently small and independent of both q and ε such that $\mu_{k_\star,q}^{\varepsilon} > 0$ for any $\varepsilon \leqslant k_\star$.*

Lemma 7.18: *There holds $\mu_{k,q}^{\varepsilon} \to -\infty$ as $k \to +\infty$ if $\sup_M f > 0$.*

Proofs for Lemmas 7.17 and 7.18 above can be found in [Ngo12, Subsection 4.2]. In the case $\sup_M f \leqslant 0$, instead of decaying to $-\infty$ as $k \to +\infty$, we obtain $\mu_{k,q}^{\varepsilon} \to +\infty$ as $k \to +\infty$ as shown in Fig. 5.

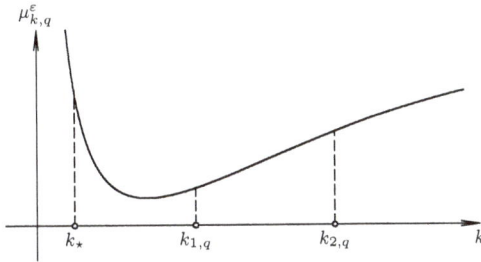

Fig. 5. The asymptotic behavior of $\mu_{k,q}^{\varepsilon}$ when $\sup_M f \leqslant 0$.

7.3.5. *The way to find solutions when either f changes sign or* $\sup_M f \leqslant 0$

Once we have the asymptotic behavior of $\mu_{k,q}^{\varepsilon}$ when k is large and small, we can start to look for solutions of Eq. (6.3) by making the curve $k \mapsto \mu_{k,q}^{\varepsilon}$ look like in Fig. 6.

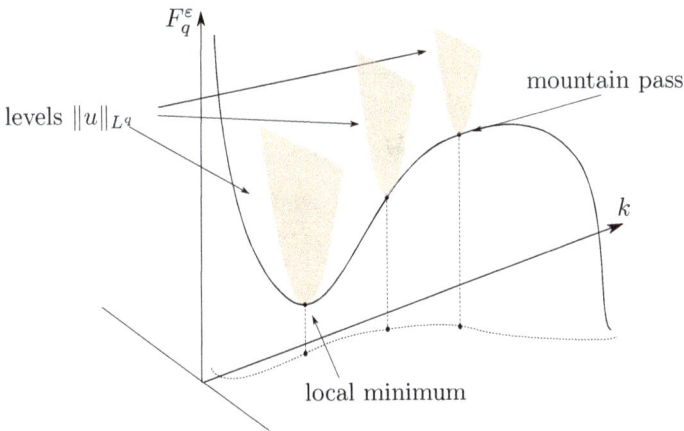

Fig. 6. Solutions will be found as minimum and mountain-pass points.

The trick is to find $k_{1,q}$ and $k_{2,q}$ in such a way that $\mu_{k_{1,q},q}^{\varepsilon} < \min\{\mu_{k_\star,q}^{\varepsilon}, \mu_{k_{2,q},q}^{\varepsilon}\}$. Then we can use standard variational techniques to look for solutions as indicated in Fig. 6.

7.3.6. *The case $\sup_M > 0$ and $\inf_M a = 0$ when $h = 0$*

Under this context, making use of the method of sub- and super-solutions is the key argument. Thanks to [Heb09], from that we learn this approach.

However, it is worth mentioning that our construction of sub-solutions is different from that of [Heb09]. We let $\varepsilon_0 > 0$ sufficiently small and then fix it so that the following inequality

$$\int_M a\, d\text{vol}_g + \varepsilon_0 < \frac{\lambda^n}{4(\eta_0)^{n-2}} \left(\frac{2n}{n-2}\right)^{n-2} \left(\int_M |f^-|\, d\text{vol}_g\right)^{1-n} \qquad (7.24)$$

still holds. Since the manifold M has unit volume, we can conclude that from (7.24), the function $a + \varepsilon_0$ verifies all assumptions in the previous subsection, thus showing that there exists a positive smooth function $\overline{u}$ solving the following equation

$$-\Delta_g \overline{u} = f\overline{u}^{\frac{n+2}{n-2}} + (a + \varepsilon_0)\overline{u}^{-\frac{3n-2}{n-2}}.$$

Obviously, $\overline{u}$ is a super-solution to (7.6), that is

$$-\Delta_g \overline{u} \geqslant f\overline{u}^{\frac{n+2}{n-2}} + a\overline{u}^{-\frac{3n-2}{n-2}}.$$

Our aim is to find a sub-solution to (7.6). Indeed, since

$$\int_M \left(a + \frac{\int_M a\, d\text{vol}_g}{\int_M |f^-|\, d\text{vol}_g} f^-\right) d\text{vol}_g = 0,$$

there exists a function $u_0 \in H^1(M)$ solving

$$-\Delta_g u_0 = a + \frac{\int_M a\, d\text{vol}_g}{\int_M |f^-|\, d\text{vol}_g} f^-. \qquad (7.25)$$

Since the right hand side of (7.25) is of class $L^p(M)$ for any $p < +\infty$, the Caldéron-Zygmund inequality tells us that the solution u_0 is of class $W^{2,p}(M)$ for any $p < +\infty$. Thanks to the Sobolev Embedding theorem [Aub98, 2.10], we can conclude that $u_0 \in C^{0,\alpha}(M)$ for some $\alpha \in (0, 1)$. In particular, the solution u_0 is continuous. Therefore, by adding a sufficiently large constant C to the function u_0 if necessary, we can always assume that $\min_M u_0 > 1$. We now find the sub-solution $\underline{u}$ of the form εu_0 for small $\varepsilon > 0$ to be determined. To this end, we first write

$$-\Delta_g \underline{u} = \varepsilon a + \frac{\varepsilon \int_M a\, d\text{vol}_g}{\int_M |f^-|\, d\text{vol}_g} f^-. \qquad (7.26)$$

Since $\max_M u_0 < +\infty$, it is easy to see that, for any $0 < \varepsilon \leqslant (\max_M u_0)^{-\frac{3n-2}{4n-4}}$, we immediately have

$$\varepsilon a \leqslant a\varepsilon^{-\frac{3n-2}{n-2}} u_0^{-\frac{3n-2}{n-2}}. \qquad (7.27)$$

Besides, since $f^- \leqslant 0$ and $\frac{2n}{n-2} > 2$, it is not difficult to see that the following inequality

$$\frac{\varepsilon \int_M a \, d\mathrm{vol}_g}{\int_M |f^-| \, d\mathrm{vol}_g} f^- \leqslant \varepsilon^{\frac{n+2}{n-2}} u_0^{\frac{n+2}{n-2}} f^-$$

holds provided

$$\varepsilon \leqslant \left(\frac{\int_M a \, d\mathrm{vol}_g}{\int_M |f^-| \, d\mathrm{vol}_g} \right)^{\frac{n-2}{4}} (\max_M u_0)^{-\frac{n^2-4}{4}}. \tag{7.28}$$

In particular, the following

$$\frac{\varepsilon \int_M a \, d\mathrm{vol}_g}{\int_M |f^-| \, d\mathrm{vol}_g} f^- \leqslant \varepsilon^{\frac{n+2}{n-2}} u_0^{\frac{n+2}{n-2}} f \tag{7.29}$$

holds provided (7.28) holds. Combining (7.26), (7.27), and (7.29), we conclude that for small ε

$$-\Delta_g \underline{u} \leqslant \varepsilon^{\frac{n+2}{n-2}} u_0^{\frac{n+2}{n-2}} f + a\varepsilon^{-\frac{3n-2}{n-2}} u_0^{-\frac{3n-2}{n-2}}.$$

In other words, we have showed that

$$-\Delta_g \underline{u} \leqslant f \underline{u}^{\frac{n+2}{n-2}} + a\underline{u}^{-\frac{3n-2}{n-2}}.$$

Finally, since $\bar{u}$ has a strictly positive lower bound, we can choose $\varepsilon > 0$ sufficiently small such that $\underline{u} \leqslant \bar{u}$. Using the sub- and super-solutions method, see [KW75, Lemma 2.6], we can conclude the existence of a positive solution u to (7.6). By a regularity result developed in [KW75], we know that u is smooth.

7.3.7. *Small of* $\sup_M a$ *can control everything in the case* $h = 0$: *Proof of Theorem 7.4*

The proof we provide here is based on the method of the sub- and super-solutions, see [KW75, Jun94].

We first construct a positive super-solution $\bar{u}$ for (7.6). By using the change of variable $u = e^v$, we get that

$$\Delta u + f u^{\frac{n+2}{n-2}} + au^{-\frac{3n-2}{n-2}} = e^v(\Delta v + |\nabla v|^2) + f e^{\frac{n+2}{n-2}v} + ae^{-\frac{3n-2}{n-2}v}.$$

Hence, it suffices to find v satisfying

$$\Delta v + |\nabla v|^2 + f e^{\frac{4}{n-2}v} + ae^{-\frac{4n-4}{n-2}v} \leqslant 0.$$

In order to do this, thanks to $\int_M f\,dvol_g < 0$, we can pick $b > 0$ small enough such that

$$|e^{\frac{4}{n-2}b\varphi} - 1| \leqslant -\frac{1}{4\sup_M f}\int_M f\,dvol_g$$

and

$$b|\nabla\varphi|^2 < -\frac{1}{4}\int_M f\,dvol_g,$$

where φ is a positive smooth solution of the following equation

$$\Delta\varphi = \int_M f\,dvol_g - f.$$

We now find the function v of the form

$$v = b\varphi + \frac{n-2}{4}\log b.$$

Indeed, by calculations, we have

$$\Delta v + |\nabla v|^2 + fe^{\frac{4}{n-2}v} + ae^{-\frac{4n-4}{n-2}v}$$

$$= \Delta\left(b\varphi + \frac{n-2}{4}\log b\right) + \left|\nabla\left(b\varphi + \frac{n-2}{4}\log b\right)\right|^2$$

$$+ fe^{\frac{4}{n-2}\left(b\varphi + \frac{n-2}{4}\log b\right)} + ae^{-\frac{4n-4}{n-2}\left(b\varphi + \frac{n-2}{4}\log b\right)}$$

$$\leqslant \frac{b}{2}\int_M f\,dvol_g + ae^{-\frac{4n-4}{n-2}b\varphi}b^{1-n}.$$

Therefore, if we assume that the function a verifies the following estimate

$$\sup_M a < -\frac{b^n}{4}e^{\frac{4n-4}{n-2}b\varphi}\int_M f\,dvol_g, \tag{7.30}$$

we then get that

$$\Delta v + |\nabla v|^2 + fe^{\frac{4}{n-2}v} + ae^{-\frac{4n-4}{n-2}v} \leqslant \frac{b}{4}\int_M f\,dvol_g < 0,$$

which concludes the existence of a super-solution $\overline{u}$. We now turn to the existence of a sub-solution. Before doing so, we can easily check that

$$\overline{u} = e^{b\varphi + \frac{n-2}{4}\log b} > b^{\frac{n-2}{4}}.$$

Since $\overline{u}$ has a strictly positive lower bound and thanks to the second stage of the proof of Theorem 7.3, we can easily construct a sup-solution $\underline{u}$ with $\underline{u} < \overline{u}$. It is important to note that the existence of a sub-solution depends heavily on the conditions $a \geqslant 0$ and $a \not\equiv 0$; and here is the only place we make use of that fact in the proof. The proof of the theorem is now complete.

7.3.8. *The case* $\sup_M f > 0$ *and* $\inf_M a = 0$ *when* $h > 0$

Under this context, making use of the method of sub- and super-solutions is the key argument. Thanks to [Heb09], from that we learn this approach. However, it is worth mentioning that our construction of sub-solutions is different from that of [Heb09]. We let $\varepsilon_0 > 0$ sufficiently small and then fix it so that the following inequality

$$\int_M a \, d\text{vol}_g + \varepsilon_0 < \frac{(2n-1)^{n-1}}{2^{2n-1}n^n} \frac{S_h}{\tau} \left(\frac{S_h \tau}{\tau \sup_M f - \int_M f \, d\text{vol}_g} \right)^{n-1} \quad (7.31)$$

still holds. Since the manifold M has unit volume, we can conclude that from (7.31), the function $a + \varepsilon_0$ verifies all assumptions in the previous subsection, thus showing that there exists a positive smooth function $\bar{u}$ solving the following equation

$$-\Delta_g \bar{u} + h\bar{u} = f\bar{u}^{\frac{n+2}{n-2}} + (a+\varepsilon_0)\bar{u}^{-\frac{3n-2}{n-2}}.$$

Obviously, $\bar{u}$ is a super-solution to (6.3), that is

$$-\Delta_g \bar{u} + h\bar{u} \geqslant f\bar{u}^{\frac{n+2}{n-2}} + a\bar{u}^{-\frac{3n-2}{n-2}}.$$

Our aim is to find a sub-solution to (6.3). In this context, we consider the following equation

$$-\Delta_g u + (h - f^-)u = a. \quad (7.32)$$

Since $h - f^- > 0$, $a \geqslant 0$, $a \not\equiv 0$, and the manifold M is compact without boundary, the standard argument shows that (7.32) always admits a weak solution, say u_0. By a standard regularity result, one can easily deduce that u_0 is at least continuous. Thus, by the Maximum Principle, we conclude $u_0 > 0$.

As before, we now find the sub-solution $\underline{u}$ of the form εu_0 for small $\varepsilon > 0$ to be determined. To this purpose, we first write

$$-\Delta_g \underline{u} + h\underline{u} = \varepsilon a + f^- \underline{u}. \quad (7.33)$$

Since $\max_M u_0 < +\infty$, it is easy to see that, for any $0 < \varepsilon \leqslant (\max_M u_0)^{-\frac{3n-2}{4n-4}}$, we immediately have

$$\varepsilon a \leqslant a\varepsilon^{-\frac{3n-2}{n-2}} u_0^{-\frac{3n-2}{n-2}}. \quad (7.34)$$

Besides, since $f^- \leqslant 0$ and $\frac{2n}{n-2} > 2$, it is not difficult to see that the following inequality

$$\varepsilon u_0 f^- \leqslant \varepsilon^{\frac{n+2}{n-2}} u_0^{\frac{n+2}{n-2}} f^-$$

holds provided $\varepsilon \leqslant (\max_M u_0)^{-1}$. In particular, the following

$$\varepsilon u_0 f^- \leqslant \varepsilon^{\frac{n+2}{n-2}} u_0^{\frac{n+2}{n-2}} f \tag{7.35}$$

holds provided $\varepsilon \leqslant (\max_M u_0)^{-1}$. Combining all estimates (7.33), (7.34), and (7.35) above, we conclude that for small ε, there holds

$$-\Delta_g \underline{u} + h\underline{u} \leqslant \varepsilon^{\frac{n+2}{n-2}} u_0^{\frac{n+2}{n-2}} f + a\varepsilon^{-\frac{3n-2}{n-2}} u_0^{-\frac{3n-2}{n-2}}.$$

In other words, we have shown that $\underline{u}$ is a sub-solution of (6.3). Finally, since $\bar{u}$ has a strictly positive lower bound, we can choose $\varepsilon > 0$ sufficiently small such that $\underline{u} \leqslant \bar{u}$. Using the sub- and super-solutions method, see [KW75, Lemma 2.6], we can conclude the existence of a positive solution u to (6.3). By a regularity result developed in [KW75], we know that u is smooth.

7.3.9. *Small of* $\sup_M f$ *can control everything in the case* $h = 0$: *Proof of Theorem 7.7*

In order to prove Theorem 7.8, we need to show that the condition (7.12) is fulfilled. Although we have not assumed any upper bound for $\int_M a \, dvol_g$, we are able to show that we can recover the condition (7.12) provided $\sup_M f$ is sufficiently small.

As always, we first assume $\inf_M a > 0$. Depending on the sign of $\int_M f \, dvol_g$, we have two cases.

Case 1. Suppose $\int_M f \, dvol_g \geqslant 0$. In this context, we can easily verify that

$$\frac{S_h}{\sup_M f} \leqslant \frac{S_h \tau}{\tau \sup_M f - \int_M f \, dvol_g}.$$

Therefore, it suffices to show that

$$\int_M a \, dvol_g < \frac{(2n-1)^{n-1}}{2^{2n-1} n^n} \frac{S_h}{\tau} \left(\frac{S_h}{\sup_M f} \right)^{n-1},$$

which is equivalent to

$$\sup_M f < \left(\frac{(2n-1)^{n-1}}{2^{2n-1} n^n} \frac{S_h^n}{\tau \int_M a \, dvol_g} \right)^{\frac{1}{n-1}}.$$

Case 2. Suppose $\int_M f \, dvol_g < 0$. In this context, we assume for a moment that $\sup_M f > 0$ is small in such a way that we can select

$$\tau = \frac{1}{\sup_M f} > \max\left\{ 1, \left(\frac{2}{S_h} \int_M h \, dvol_g \right)^{\frac{n}{n-2}} \right\}.$$

Then, thanks to $f = f^+ + f^-$, we have

$$\frac{S_h \tau}{\tau \sup_M f - \int_M f \, dvol_g} = \frac{1}{\sup_M f} \frac{S_h}{1 - \int_M f \, dvol_g}$$

$$\geq \frac{1}{\sup_M f} \frac{S_h}{1 + \int_M |f^-| \, dvol_g}.$$

Therefore, it suffices to show that

$$\int_M a \, dvol_g < \frac{(2n-1)^{n-1}}{2^{2n-1} n^n} S_h \sup_M f \left(\frac{1}{\sup_M f} \frac{S_h}{1 + \int_M |f^-| \, dvol_g} \right)^{n-1},$$

which is equivalent to

$$\sup_M f < \left(\frac{(2n-1)^{n-1}}{2^{2n-1} n^n} \frac{S_h^n}{(1 + \int_M |f^-| \, dvol_g) \int_M a \, dvol_g} \right)^{\frac{1}{n-2}}.$$

From our calculation above, we conclude that there exists some positive constant $C > 0$ depending only on a, h, and f^- such that if $0 < \sup_M f < C$, our Eq. (6.3) always admits at least one positive smooth solution.

It remains to consider the case $\inf_M a = 0$. However, since the size of a plays no role in the above calculation, we can freely add a small constant ε_0 to a as in the second stage of the proof of Theorem 7.6. This procedure ensures that we always get a super-solution of (6.3) with a strictly positive lower bound and this is enough since a suitable positive sub-solution always exists.

As mentioned at the beginning of the section, we are considering the case when M is closed and τ is constant everywhere in M. The cases when M is either non-compact or compact with boundary as well as non-constant τ are very delicate. However, due to the length of the notes, we omit it.

8. Other Result: A Liouville Type Result

In the last part of this notes, we study a Liouville type result for positive, smooth solutions of (6.3). For the sake of simplicity, we only consider Eq. (6.3) in the case when h, f, and a are constants with, of course, $a > 0$. Our aim here is to give some sufficient conditions so that (6.3) has only constant solution.

Note that by a scaling argument we may assume $a = 1$; hence, in this section, we are interested in the following model equation

$$-\Delta_g u + hu = \lambda u^q + u^{-q-2}, \quad u > 0, \tag{8.1}$$

where h and λ are constants. It is worth noticing that after using a scaling the sign of h still matches the sign of the Yamabe-scalar field conformal invariant since $a > 0$. We also notice that the exponent $q > 0$ here is arbitrary rather than the number $\frac{n+2}{n-2}$ and will be specified later. Besides, we could also consider the case $q < 0$ if physical problems motivate it; however, we shall not touch the case $q \leqslant 0$ in this section.

Our result was inspired by a couple of recent papers [MX09, Ma10]. In these papers, the authors considered the following model equation

$$- \Delta_g u = -u^q + u^{-q-2}, \quad u > 0, \tag{8.2}$$

in $\mathbb{R}^n$ with the standard metric where $q \in (1, \frac{2n}{n-2})$, that is, $h = 0$ and $\lambda = -1$. First, they proved in [MX09] that smooth positive solutions of (8.2) are uniformly bounded. Then by using the idea from Redheffer [Red60], Ma [Ma10, Theorem 1] was able to prove that any smooth positive solution of (8.2) is constant; hence, is equal to 1. In [Bre11], Brezis used a different approach to establish, among other things, such a Liouville type result. Besides, it was shown in [Ma10, Theorem 2] that the similar Liouville type result is also true for smooth positive solutions for (8.2) in a complete non-compact Riemannian manifold with the Ricci curvature bounded from below.

Motivated by all discussion above, in this section we prove that any smooth positive solution of Eq. (8.1) in a complete compact Riemannian manifold with the Ricci curvature bounded from, whose the bound will be determined, is constant. To be precise, we now state our main result.

Theorem 8.1: *Let (M, g) be a smooth closed Riemannian manifold of dimension $n \geqslant 3$. Let h, λ, and $q > 0$ be constants. Then there is a constant $K(n, q, h)$ depending only on n, q and h so that if $\mathrm{Ric}_g \geqslant K$ in the sense of quadratic forms, then every smooth positive solution of (8.1) is constant provided that in the case $h > 0$, $\lambda > 0$, we have to restrict $q \leqslant \frac{n+2}{n-2}$.*

Our result can be formulated as in Table 5. Surprisingly enough, as can be seen from Table 5, the constant K does not depend on λ. It is worth noticing that by integrating both sides of (8.2) over M, one easily gets that Eq. (8.2) has no positive solution if $h \leqslant 0$ and $\lambda \geqslant 0$. The papers by Gidas–Spruck [GS81] and M. Bidaut-Véron–L. Véron [BVV91] were sources of inspiration.

Finally, as usual, we should mention that the content of this section was adapted from [NX11]. We now prove Theorem 8.1 in detail.

Table 5. The Liouville type result in terms of the Ricci curvature.

h	λ	q	$\mathrm{Ric}_g \geqslant K$
$h < 0$	$\lambda < 0$	$q \geqslant 1$	$\mathrm{Ric}_g \geqslant 0$
		$q < 1$	$\mathrm{Ric}_g \geqslant \frac{n-1}{n}(q-1)h$
$h = 0$	$\lambda < 0$	$q > 0$	$\mathrm{Ric}_g \geqslant 0$
$h > 0$	$\lambda < 0$	$q > 0$	$\mathrm{Ric}_g > -\frac{n-1}{n}h$
	$\lambda > 0$	$q \leqslant \frac{n+2}{n-2}$	$\mathrm{Ric}_g \geqslant \frac{n-1}{n}(q-1)h$

8.1. *Some basic computations*

For simplicity, we shall use $\int_M$ to denote the integral with respect to the measure induced by the metric g, that is $\int_M = \int_M d\mathrm{vol}_g$. First, we need a preparation. For some $\beta \neq 0$ to be determined later, we denote $u = v^{-\beta}$. A direct computation shows that

$$\Delta_g v = (\beta + 1)\frac{|\nabla v|^2}{v} + \frac{1}{\beta}(-\Delta_g u)v^{\beta+1}.$$

This and the fact that

$$-\Delta_g u = -hv^{-\beta} + \lambda v^{-\beta q} + v^{\beta(q+2)}$$

give

$$\Delta_g v = (\beta + 1)\frac{|\nabla v|^2}{v} + \frac{1}{\beta}(-hv + \lambda v^{1-\beta(q-1)} + v^{1+\beta(q+3)}). \qquad (8.3)$$

Applying the well-known Bochner-Lichnerowicz-Weitzenböck formula [Aub98] to function v, one obtains

$$\frac{1}{2}\Delta_g(|\nabla v|^2) = |\mathrm{Hess}(v)|^2 + \langle \nabla \Delta_g v, \nabla v \rangle + \mathrm{Ric}_g(\nabla v, \nabla v). \qquad (8.4)$$

Multiply both sides of (8.4) by v^γ where $\gamma \in \mathbb{R}$ will be chosen later and integrate on M, to obtain

$$A + B + C + D = 0, \qquad (8.5)$$

where

$$A = \frac{1}{n}\int_M v^\gamma(\Delta_g v)^2 + \int_M v^\gamma\left(|\mathrm{Hess}(v)|^2 - \frac{1}{n}(\Delta_g v)^2\right), \qquad (8.6)$$

$$B = \int_M v^\gamma \langle \nabla \Delta_g v, \nabla v \rangle,$$

$$C = -\frac{1}{2}\int_M v^\gamma \Delta_g(|\nabla v|^2),$$

and

$$D = \int_M v^\gamma \mathrm{Ric}_g(\nabla v, \nabla v).$$

We notice that γ may not necessarily be nonzero. Besides, there holds

$$J = \int_M v^\gamma \left(|\mathrm{Hess}(v)|^2 - \frac{1}{n}(\Delta_g v)^2\right) \geqslant 0$$

since it is well-known that

$$|\mathrm{Hess}(v)|^2 - \frac{1}{n}(\Delta_g v)^2 \geqslant 0.$$

8.2. *Computations of A, B, and C*

We treat the first term of A in the following way. In fact, using (8.3), one obtains

$$\frac{1}{n}\int_M v^\gamma(\Delta_g v)^2$$

$$= \frac{1}{n}\int_M v^\gamma(\Delta_g v)\left((\beta+1)\frac{|\nabla v|^2}{v} + \frac{1}{\beta}(-hv + \lambda v^{1-\beta(q-1)} + v^{1+\beta(q+3)})\right)$$

$$= \frac{\beta+1}{n}\int_M v^{\gamma-1}|\nabla v|^2\left((\beta+1)\frac{|\nabla v|^2}{v}\right)$$

$$+ \frac{\beta+1}{n}\int_M v^{\gamma-1}|\nabla v|^2\left(\frac{1}{\beta}(-hv + \lambda v^{1-\beta(q-1)} + v^{1+\beta(q+3)})\right)$$

$$+ \frac{1}{n\beta}\int_M (-hv^{\gamma+1} + \lambda v^{\gamma+1-\beta(q-1)} + v^{\gamma+1+\beta(q+3)})(\Delta_g v)$$

$$= \frac{(\beta+1)^2}{n}\int_M v^{\gamma-2}|\nabla v|^4$$

$$+ \frac{\beta+1}{n\beta}\int_M |\nabla v|^2(-hv^\gamma + \lambda v^{\gamma-\beta(q-1)} + v^{\gamma+\beta(q+3)})$$

$$+ \frac{1}{n\beta}\int_M (-hv^{\gamma+1} + \lambda v^{\gamma+1-\beta(q-1)} + v^{\gamma+1+\beta(q+3)})(\Delta_g v).$$

Therefore, this and (8.6) imply that

$$A = J + \frac{(\beta+1)^2}{n} \int_M v^{\gamma-2}|\nabla v|^4$$
$$+ \frac{\beta+1}{n\beta} \int_M |\nabla v|^2(-hv^\gamma + \lambda v^{\gamma-\beta(q-1)} + v^{\gamma+\beta(q+3)}) \qquad (8.7)$$
$$+ \frac{1}{n\beta} \int_M (-hv^{\gamma+1} + \lambda v^{\gamma+1-\beta(q-1)} + v^{\gamma+1+\beta(q+3)})(\Delta_g v).$$

By the divergence theorem, it holds

$$\int_M v^{\gamma+1}\Delta_g v = -(\gamma+1)\int_M v^\gamma|\nabla v|^2,$$
$$\int_M v^{\gamma+1-\beta(q-1)}\Delta_g v = -(\gamma+1-\beta(q-1))\int_M v^{\gamma-\beta(q-1)}|\nabla v|^2,$$
$$\int_M v^{\gamma+1+\beta(q+3)}\Delta_g v = -(\gamma+1+\beta(q+3))\int_M v^{\gamma+\beta(q+3)}|\nabla v|^2.$$

Therefore, we can further simplify (8.7) as follows

$$A = J + \frac{(\beta+1)^2}{n} \int_M v^{\gamma-2}|\nabla v|^4$$
$$+ \frac{\beta+1}{n\beta} \int_M |\nabla v|^2(-hv^\gamma + \lambda v^{\gamma-\beta(q-1)} + v^{\gamma+\beta(q+3)})$$
$$- \frac{\lambda}{n\beta}(\gamma+1-\beta(q-1)) \int_M v^{\gamma-\beta(q-1)}|\nabla v|^2$$
$$+ \frac{1}{n\beta}(\gamma+1) \int_M hv^\gamma|\nabla v|^2 - \frac{1}{n\beta}(\gamma+1+\beta(q+3)) \int_M v^{\gamma+\beta(q+3)}|\nabla v|^2.$$

Thus, finally, we have

$$A = J + \frac{(\beta+1)^2}{n} \int_M v^{\gamma-2}|\nabla v|^4 + \frac{h(\gamma-\beta)}{n\beta} \int_M v^\gamma|\nabla v|^2$$
$$- \frac{\lambda(\gamma-\beta q)}{n\beta} \int_M v^{\gamma-\beta(q-1)}|\nabla v|^2 \qquad (8.8)$$
$$- \frac{\gamma+\beta(q+2)}{n\beta} \int_M v^{\gamma+\beta(q+3)}|\nabla v|^2.$$

For the term B, again using (8.3), we have

$$B = \int_M v^\gamma \left\langle \nabla\left((\beta+1)\frac{|\nabla v|^2}{v} + \frac{1}{\beta}(-hv + \lambda v^{1-\beta(q-1)} + v^{1+\beta(q+3)})\right), \nabla v \right\rangle$$
$$= (\beta+1)\int_M \left\langle \nabla\left(\frac{|\nabla v|^2}{v}\right), v^\gamma\nabla v \right\rangle$$

$$- \frac{h}{\beta} \int_M v^\gamma |\nabla v|^2 + \lambda \frac{1 - \beta(q-1)}{\beta} \int_M v^{\gamma - \beta(q-1)} |\nabla v|^2$$
$$+ \frac{1 + \beta(q+3)}{\beta} \int_M v^{\gamma + \beta(q+3)} |\nabla v|^2.$$

Notice that

$$\int_M \left\langle \nabla \left(\frac{|\nabla v|^2}{v} \right), v^\gamma \nabla v \right\rangle = - \int_M \frac{|\nabla v|^2}{v} \nabla \cdot (v^\gamma \nabla v)$$
$$= -\gamma \int_M v^{\gamma - 2} |\nabla v|^4 - \int_M v^{\gamma - 1} |\nabla v|^2 \Delta_g v.$$

Therefore,

$$B = - (\beta + 1)\gamma \int_M v^{\gamma - 2} |\nabla v|^4 - (\beta + 1) \int_M v^{\gamma - 1} |\nabla v|^2 \Delta_g v$$
$$- \frac{h}{\beta} \int_M v^\gamma |\nabla v|^2 + \lambda \frac{1 - \beta(q-1)}{\beta} \int_M v^{\gamma - \beta(q-1)} |\nabla v|^2$$
$$+ \frac{1 + \beta(q+3)}{\beta} \int_M v^{\gamma + \beta(q+3)} |\nabla v|^2.$$

Again we use (8.3) to reach at

$$B = -(\beta + 1)\gamma \int_M v^{\gamma - 2} |\nabla v|^4$$
$$- (\beta + 1) \int_M v^{\gamma - 1} |\nabla v|^2 \left((\beta + 1) \frac{|\nabla v|^2}{v} \right)$$
$$- (\beta + 1) \int_M v^{\gamma - 1} |\nabla v|^2 \left(\frac{1}{\beta}(-hv + \lambda v^{1 - \beta(q-1)} + v^{1 + \beta(q+3)}) \right)$$
$$- \frac{h}{\beta} \int_M v^\gamma |\nabla v|^2 + \lambda \frac{1 - \beta(q-1)}{\beta} \int_M v^{\gamma - \beta(q-1)} |\nabla v|^2$$
$$+ \frac{1 + \beta(q+3)}{\beta} \int_M v^{\gamma + \beta(q+3)} |\nabla v|^2.$$

By simplifying the right hand side of the above identity, one gets

$$B = - (\beta + 1)(\gamma + \beta + 1) \int_M v^{\gamma - 2} |\nabla v|^4$$
$$+ h \int_M v^\gamma |\nabla v|^2 - \lambda q \int_M v^{\gamma - \beta(q-1)} |\nabla v|^2 \tag{8.9}$$
$$+ (q + 2) \int_M v^{\gamma + \beta(q+3)} |\nabla v|^2.$$

For the term C, we first observe that

$$\int_M v^\gamma \Delta_g (|\nabla v|^2) = \int_M \Delta_g (v^\gamma) |\nabla v|^2.$$

Therefore,

$$
\begin{aligned}
C &= -\frac{1}{2}\int_M |\nabla v|^2 \Delta_g(v^\gamma) \\
&= -\frac{1}{2}\int_M |\nabla v|^2\left(\gamma v^{\gamma-1}\Delta_g v + \gamma(\gamma-1)v^{\gamma-2}|\nabla v|^2\right) \\
&= -\frac{1}{2}\int_M \gamma v^{\gamma-1}|\nabla v|^2\left((\beta+1)\frac{|\nabla v|^2}{v}\right) \\
&\quad -\frac{1}{2}\int_M \gamma v^{\gamma-1}|\nabla v|^2\left(\frac{1}{\beta}(-hv + \lambda v^{1-\beta(q-1)} + v^{1+\beta(q+3)})\right) \\
&\quad -\frac{\gamma(\gamma-1)}{2}\int_M v^{\gamma-2}|\nabla v|^4.
\end{aligned}
$$

In other words,

$$
\begin{aligned}
C &= \frac{\gamma h}{2\beta}\int_M v^\gamma|\nabla v|^2 - \frac{\gamma\lambda}{2\beta}\int_M v^{\gamma-\beta(q-1)}|\nabla v|^2 \\
&\quad -\frac{\gamma}{2\beta}\int_M v^{\gamma+\beta(q+3)}|\nabla v|^2 - \frac{\gamma(\gamma+\beta)}{2}\int_M v^{\gamma-2}|\nabla v|^4.
\end{aligned}
\tag{8.10}
$$

We now have enough information to treat (8.5).

8.3. *The transformed equation*

By using (8.8)-(8.10), one can see that (8.5) reduces to

$$
\begin{aligned}
&J + \left(\frac{(\beta+1)^2}{n} - (\beta+1)(\gamma+\beta+1) - \frac{\gamma(\gamma+\beta)}{2}\right)\int_M v^{\gamma-2}|\nabla v|^4 \\
&+ \left(\frac{h(\gamma-\beta)}{n\beta} + h + \frac{\gamma h}{2\beta}\right)\int_M v^\gamma|\nabla v|^2 \\
&+ \left(-\frac{\lambda(\gamma-\beta q)}{n\beta} - \lambda q - \frac{\gamma\lambda}{2\beta}\right)\int_M v^{\gamma-\beta(q-1)}|\nabla v|^2 \\
&+ \left(-\frac{\gamma+\beta(q+2)}{n\beta} + (q+2) - \frac{\gamma}{2\beta}\right)\int_M v^{\gamma+\beta(q+3)}|\nabla v|^2 \\
&+ \int_M v^\gamma \mathrm{Ric}_g(\nabla v, \nabla v) = 0.
\end{aligned}
\tag{8.11}
$$

For simplicity, we rewrite Eq. (8.11) as

$$
\begin{aligned}
&J + a\int_M v^{\gamma-2}|\nabla v|^4 + \int_M v^\gamma\left(c|\nabla v|^2 + \mathrm{Ric}_g(\nabla v, \nabla v)\right) \\
&+ b\int_M v^{\gamma-\beta(q-1)}|\nabla v|^2 + d\int_M v^{\gamma+\beta(q+3)}|\nabla v|^2 = 0,
\end{aligned}
$$

where

$$a = -\frac{1}{2}\left(\gamma^2 + (3\beta + 2)\gamma + 2\frac{n-1}{n}(\beta+1)^2\right),$$

$$b = \frac{\lambda(n+2)}{2n}\left(-\frac{\gamma}{\beta} - 2q\frac{n-1}{n+2}\right),$$

$$c = \frac{n+2}{2n}\left(\frac{\gamma}{\beta} + 2\frac{n-1}{n+2}\right)h,$$

$$d = \frac{n+2}{2n}\left((q+2)\frac{2(n-1)}{n+2} - \frac{\gamma}{\beta}\right).$$

Next we wish to describe the method used in the present paper. Our goal was to find $\beta \neq 0$ and $\gamma \in \mathbb{R}$ such that

$$a \geqslant 0, \quad b \geqslant 0, \quad \text{Ric}_g + cg \geqslant 0, \quad d \geqslant 0. \tag{8.12}$$

Having (8.12), our result follows easily since $|\nabla v| = 0$ since $a \geqslant 0$ forces $d > 0$. Thus, the key point is $a \geqslant 0$. To better serve this purpose, we set $y = 1 + \frac{1}{\beta}$ and $\delta = -\frac{\gamma}{\beta}$ where $y \neq 1$ and $\delta \in \mathbb{R}$. Thus the set of conditions in (8.12) becomes

$$2\frac{n-1}{n}y^2 - 2\delta y + \delta^2 - \delta \leqslant 0, \tag{8.13}$$

$$\lambda\left(\delta - 2q\frac{n-1}{n+2}\right) \geqslant 0, \tag{8.14}$$

$$\frac{2n}{n+2}\text{Ric}_g \geqslant h\left(\delta - 2\frac{n-1}{n+2}\right)g, \tag{8.15}$$

and

$$\delta \geqslant -2(q+2)\frac{n-1}{n+2}. \tag{8.16}$$

In view of (8.13), it is necessary to have

$$\frac{\delta}{n}(2(n-1) - (n-2)\delta) \geqslant 0$$

which is equivalent to

$$0 \leqslant \delta \leqslant \frac{2(n-1)}{n-2}. \tag{8.17}$$

With (8.17) in hand, one can see that (8.16) is automatically satisfied. Moreover, $d > 0$ provided $\delta \geqslant 0$. Thus, our set of conditions now reduces to (8.14), (8.15), and (8.17). Notice that if inequalities in (8.17) are strict, then we can always find some $y \neq 1$ verifying (8.13).

8.4. *Proof of Theorem 8.1*

For the sake of clarity, we split our studying into four cases depending on the sign of h and λ.

8.4.1. *The case $h < 0$*

In this case, it is necessary to have $\lambda < 0$. Then the condition (8.14) and the lower bound for δ in (8.17) imply

$$0 \leqslant \delta \leqslant 2q \frac{n-1}{n+2}. \tag{8.18}$$

Combining (8.17) and (8.18) gives

$$0 \leqslant \delta \leqslant \min \left\{ 2q \frac{n-1}{n+2}, \frac{2(n-1)}{n-2} \right\}. \tag{8.19}$$

There are two possible sub-cases.

Case 1. Suppose $q \geqslant \frac{n+2}{n-2}$. Then we claim that, with $K = 0$, we can always select δ such that it satisfies (8.14), (8.15), and (8.17). To this end, we notice that the right hand side of (8.15) is always non-negative. In order to see that (8.14), (8.15), and (8.17) hold, we have to select $\delta = \frac{2(n-1)}{n+2}$. Then we have to choose y such that Eq. (8.13) holds. However we are left without many choices but one, that is, $y = \frac{n}{n+2} \neq 1$. This is enough to serve our purpose since the left hand side of (8.13) equals $-\frac{8(n-1)}{(n+2)^2}$ when y is equal to $y = \frac{n}{n+2}$.

Case 2. Suppose $q < \frac{n+2}{n-2}$. Then δ needs to satisfy (8.18). With this region for δ, the right hand side of (8.15) is not smaller than $2(q-1)\frac{n-1}{n+2}hg$. Thus, we select

$$K = \frac{n-1}{n}(q-1)h.$$

Similarly as above, then we may choose $\delta = \frac{2q(n-1)}{n+2}$ to make sure that (8.15), and (8.17) hold. Now it is easy to find some $y \neq 1$ satisfying (8.13) since the solution of (8.13) is an interval. For example, when $q \neq \frac{n+2}{n}$, we may choose $y = \frac{qn}{n+2}$ since the left hand side of (8.13) now equals

$$\frac{2q(n-1)(q(n-2)-(n+2))}{(n+2)^2}.$$

Otherwise, we may choose $y = 1 - \sqrt{2/n}$, in this case, the left hand side of (8.13) now vanishes.

8.4.2. The case $h = 0$

Under this case, the right hand side of (8.15) vanishes, thus it is enough to take $K = 0$. Besides, it is necessary to have $\lambda < 0$. Therefore, δ must satisfy (8.19). It is now a simple task to find some δ and $y \neq 1$ verifying both conditions (8.14) and (8.17).

8.4.3. The case $h > 0$, $\lambda \leqslant 0$

Again, in this context, δ has to satisfy (8.19). First we show that $K = -\frac{n-1}{n}h$ is enough. Indeed, this condition can be rewritten as

$$\frac{2n}{n+2}\mathrm{Ric}_g \geqslant h\left(-2\frac{n-1}{n+2}\right)g. \tag{8.20}$$

Under the condition (8.20), we have to select $\delta = 0$. In order to see how could this choice of δ work, we just go back to (8.11) to get

$$
\begin{aligned}
J &- \frac{n-1}{n}(\beta+1)^2 \int_M \frac{|\nabla v|^4}{v^2} \\
&+ \frac{n-1}{n}h \int_M |\nabla v|^2 - \frac{n-1}{n}\lambda q \int_M v^{-\beta(q-1)}|\nabla v|^2 \\
&+ \frac{n-1}{n}(q+2)\int_M v^{\beta(q+3)}|\nabla v|^2 + \int_M \mathrm{Ric}_g(\nabla v, \nabla v) = 0.
\end{aligned} \tag{8.21}
$$

Clearly, we have no choice but $\beta = -1$ or equivalently, $y = 0$. With this choice of y, we immediately see that the left hand side of (8.21) is non-negative. This forces $\nabla v = 0$ thus giving us the desired result.

8.4.4. The case $h > 0$, $\lambda > 0$

Under this case, it follows from (8.14) and (8.17) that

$$2q\frac{n-1}{n+2} \leqslant \delta \leqslant \frac{2(n-1)}{n-2}.$$

In other words, it is necessary to have $q \leqslant \frac{n+2}{n-2}$. Our choice for K is that

$$K = \frac{n-1}{n}(q-1)h.$$

We will see how this condition is enough for our argument.

Case 1. Suppose $q < \frac{n+2}{n-2}$. We rewrite the condition for Ricci curvature in the following way

$$\frac{2n}{n+2}\mathrm{Ric}_g \geqslant h\left(2q\frac{n-1}{n+2} - 2\frac{n-1}{n+2}\right)g.$$

Thus, we may choose $\delta = \frac{2q(n-1)}{n+2}$. Consequently, the conditions (8.14) and (8.15) clearly hold. Therefore, we may select $y \neq 1$ verifying (8.13) since $\delta \in (0, \frac{2(n-1)}{n-2})$ as we have already done in the second case when $h < 0$.

Case 2. Suppose $q = \frac{n+2}{n-2}$. Then necessarily $\delta = \frac{2(n-1)}{n-2}$ which verifies (8.14). The condition for Ricci curvature can be rewritten as

$$\frac{2n}{n+2}\,\mathrm{Ric}_g \geqslant h\left(2\frac{n-1}{n-2} - 2\frac{n-1}{n+2}\right)g = \left(\frac{8h(n-1)}{n^2-4}\right)g.$$

Thus, we can pick $K = \frac{4h(n-1)}{n(n-2)}$ and clearly (8.15) holds. It suffices to find some $y \neq 1$ verifying (8.13). Due to the fact that $q = \frac{n+2}{n-2}$, we only have one choice for y, that is, $y = \frac{nq}{n+2}$. Thanks to $q = \frac{n+2}{n-2}$, we immediately see that $y = \frac{n}{n-2} \neq 1$. With this, the left hand side of (8.13) vanishes as required.

8.5. *Proof of Theorem 8.1 completed*

Finally let us assume that u is a smooth positive solution of Eq. (8.2). From our discussion above, we know that all inequalities in (8.12) are fulfilled. In fact, we have already shown that $d > 0$. Consequently,

$$\int_M v^{\gamma+\beta(q+3)}|\nabla v|^2 = 0$$

which implies that v, hence u is a constant.

Acknowledgments

The content of this unpretentious notes is based on four lectures given at IMS of the National University of Singapore during *Winter School on Scalar Curvature and Related Problems* in December 2014. The author wants to thank Professor Xingwang Xu and IMS for providing such a nice opportunity and lovely atmosphere during my stay. Thanks also go to the Vietnam Institute for Advanced Study in Mathematics (VIASM) for hosting and financial support where part of this note were writing.

References

[AP03] B. ABDELLAOUI, I. PERAL, Existence and nonexistence results for quasilinear elliptic equations involving the *p*-Laplacian with a critical potential, *Ann. Mat. Pura Appl.* **182** (2003), pp. 247–270. 184

[AR73] A. AMBROSETT, P. RABINOWITZ, Dual variational methods in critical point theory and applications, *J. Funct. Anal.* **14** (1973), pp. 349–381. 174, 182

[Aub98]   T. AUBIN, Some nonlinear problems in Riemannian geometry, Springer
           monographs in mathematics, Springer, New York, 1998. 166, 192, 199

[BI04]   R. BARTNIK, J. ISENBERG, The constraint equations, The Ein-
           stein equations and the large scale behavior of gravitational fields,
           Birkhauser, Basel, 2004, pp. 1–38. 121, 173

[BVV91]   M. BIDAUT-VÉRON, L. VÉRON, Nonlinear elliptic equations on com-
           pact Riemannian manifolds and asymptotics of Emden equations, In-
           vent. Math. **106** (1991), pp. 489–539. 198

[Bre11]   H. BREZIS, Comments on two Notes by L. Ma and X. Xu, C. R. Math.
           Acad. Sci. Paris **349** (2011), pp. 269–271. 198

[CBru09]   Y. CHOQUET-BRUHAT, General relativity and the Einstein equations,
           Oxford Mathematical Monographs, Oxford University Press, 2009. 121,
           125, 137, 151, 160, 170

[CBG69]   Y. CHOQUET-BRUHAT, R. GEROCH, Global aspects of the Cauchy
           problem in general relativity, Comm. Math. Phys. **14** 1969, pp. 329–
           335. 138

[CBIP07]   Y. CHOQUET-BRUHAT, J. ISENBERG, D. POLLACK, The constraint
           equations for the Einstein-scalar field system on compact manifolds,
           Class. Quantum Grav. **24** (2007), pp. 808–828. 166, 167, 168, 169, 170,
           171, 176, 178, 181

[CBY80]   Y. CHOQUET-BRUHAT, J.W. YORK, The Cauchy problem, General rel-
           ativity and gravitation, Vol. 1, pp. 99–172, Plenum, New York-London,
           1980. 149

[CD03]   P.T. CHRUŚCIEL, E. DELAY, On mapping properties of the general
           relativistic constraints operator in weighted function spaces, with ap-
           plications, Mém. Soc. Math. Fr. (N.S.) No. **94** (2003), vi+103 pp. 149

[CGP10]   P.T. CHRUŚCIEL, G.J. GALLOWAY, D. POLLACK, Mathematical gen-
           eral relativity: A sampler, Bull. Amer. Math. Soc. **47** (2010), pp. 567–
           638. 121

[Cor]   J. CORVINO, Introduction to general relativity and the Einstein con-
           straint equations, Lecture notes. 121

[Cor00]   J. CORVINO, Scalar curvature deformation and a gluing construction
           for the Einstein constraint equations, Comm. Math. Phys. **214** (2000),
           pp. 137–189. 149

[CS06]   J. CORVINO, R. SCHOEN, On the asymptotics for the vacuum Einstein
           constraint equations J. Differential Geom. **73** (2006), pp. 185–217. 149

[Ger70]   R. GEROCH, Domain of dependence, J. Math. Phys. **11** (1970), pp.
           437–449. 136

[GS81]   B. GIDAS, J. SPRUCK, Global and local behaviour of positive solutions
           of nonlinear elliptic equations Commun. Pure Appl. Math. **35** (1981),
           pp. 525–598. 198

[Gou12]   É. GOURGOULHON, 3 + 1 Formalism in General Relativity, Lecture
           Notes in Physics 846, 2012. 121, 125

[Heb09]   E. HEBEY, Existence, stability and instability for Einstein-scalar field
           Lichnerowicz equations, unpublished work. 191, 192, 195

[HPP09]   E. HEBEY, F. PACARD, D. POLLACK, A variational analysis of

Einstein-scalar field Lichnerowicz equations on compact Riemannian manifolds, *Commun. Math. Phys.* **278** (2008), pp. 117–132. 178, 179, 182, 186

[Ise95]   J. ISENBERG, Constant mean curvature solutions of the Einstein constraint equations on closed manifolds, *Classical Quantum Gravity* **12** (1995), pp. 2249–2274. 170, 173, 174

[Jun94]   Y.T. JUNG, On the elliptic equation $(4(n-1)/(n-2))\Delta u + K(x)u^{(n+2)/(n-2)} = 0$ and the conformal deformation of Riemannian metrics, *Indiana Univ. Math. J.* **43** (1994), pp. 7370–746. 193

[KW75]   J. KAZDAN, F. WARNER, Scalar curvature and conformal deformation of Riemannian structure, *J. Differential Geom.* **10** (1975), pp. 113–134. 193, 196

[Ler53]   J. LERAY, *Hyperbolic differential equations*, Mimeographed, IAS Princeton, 1953. 136

[Ma10]   L. MA, Liouville type theorem and uniform bound for the Lichnerowicz equation and the Ginzburg-Landau equation, *C. R. Math. Acad. Sci. Paris* **348** (2010), pp. 993–996. 198

[MW12]   L. MA, J. WEI, Stability and multiple solutions to Einstein-scalar field Lichnerowicz equation on manifolds, *J. Math. Pures App.* **99** (2013), pp. 174–186. 179

[MX09]   L. MA, X. XU, Uniform bound and a non-existence result for Lichnerowicz equation in the whole n-space, *C. R. Math. Acad. Sci. Paris* **347** (2009), pp. 805–808. 198

[Max04]   D. MAXWELL, Initial data for black holes and rough spacetimes, *Ph.D. dissertation*, University of Washington, 2004.

[Mon86]   V. MONCRIEF, Reduction of Einstein's equations for vacuum spacetimes with spacelike $U(1)$ isometry groups, *Ann. Physics* **167** (1986), pp. 118–142. 161, 164

[Mon90]   V. MONCRIEF, Reduction of the Einstein-Maxwell and Einstein-Maxwell-Higgs equations for cosmological spacetimes with spacelike $U(1)$ isometry groups, *Class. Quantum Grav.* **7** (1990), pp. 329–352. 161

[Ngo12]   Q.A. NGÔ, *The Einstein-scalar field Lichnerowicz equations on compact Riemannian manifolds*, Ph.D. dissertation, National University of Singapore, 2012. 186, 190

[NX11]   Q.A. NGÔ, X. XU, *Liouville type result for smooth positive solutions of the Einstein-scalar field Lichnerowicz equations on compact Riemannian manifolds*, unpublished notes, 2011. 198

[NX12]   Q.A. NGÔ, X. XU, Existence results for the Einstein-Scalar field Lichnerowicz equations on compact Riemannian manifolds, *Adv. Math.* **230** (2012), pp. 2378–2415. 174, 175, 181

[NX14]   Q.A. NGÔ, X. XU, Existence results for the Einstein-scalar field Lichnerowicz equations on compact Riemannian manifolds in the positive case, *Bull. Inst. Math. Acad. Sin. (N.S.)* **9** (2014), pp. 451–485. 174, 181

[NX15]   Q.A. NGÔ, X. XU, Existence results for the Einstein-scalar field Lichnerowicz equations on compact Riemannian manifolds in the null case, *Comm. Math. Phys.* **334** (2015), pp. 193–222. 174, 181

[NZ15]   Q.A. NGÔ, H. ZHANG, Prescribed Webster scalar curvature on compact CR manifolds with negative conformal invariants, *J. Diff. Equations* **258** (2015), pp. 4443–4490. 182

[Rau95]   A. RAUZY, Courbures scalaires des variétés d'invariant conforme négatif, *Trans. Amer. Math. Soc.* **347** (1995), pp. 4729–4745. 175, 183

[Red60]   R. REDHEFFER, On the inequality $\Delta u \geqq f(u, |\text{grad } u|)$, *J. Math. Anal. Appl.* **1** 1960, pp. 277–299. 198

[Str08]   M. STRUWE, Variational methods: Applications to nonlinear partial differential equations and hamiltonian systems, Springer, Berlin, 2008. 174, 182, 188

[Tay11]   M.E. TAYLOR, *Partial differential equations III. Nonlinear equations*, 2nd edition, Applied Mathematical Sciences **117**, Springer, New York, 2011. 121

[Tru68]   N. TRUDINGER, Remarks concerning the conformal deformation of Riemannian structures on compact manifolds, *Ann. Scuola Norm. Sup. Pisa (3)* **22** (1968), pp. 265–274. 166

[Yam60]   H. YAMABE, On a deformation of Riemannian structures on compact manifolds, *Osaka Math. J.* **12** (1960), pp. 21–37. 166

www.ingramcontent.com/pod-product-compliance
Lightning Source LLC
Chambersburg PA
CBHW050600190326
41458CB00007B/2119